DCN 99.803931.02

A Comparison of Gasification and Incineration of Hazardous Wastes

Report

Prepared for:

U.S. Department of Energy
National Energy Technology Laboratory (NETL)
3610 Collins Ferry Road
Morgantown, West Virginia 26505

Prepared by:

Radian International LLC
8501 North MoPac Blvd.
Austin, Texas 78759

Project Manager

Bob Wetherold, Ph.D., P.E.

Authors

Doug Orr
David Maxwell

March 30, 2000

Abstract

Gasification is a technology that has been widely used in commercial applications for more than 50 years in the production of fuels and chemicals. Current trends in the chemical manufacturing and petroleum refinery industries indicate that use of gasification facilities to produce synthesis gas ("syngas") will continue to increase. Attractive features of the technology include: 1) the ability to produce a consistent, high-quality syngas product that can be used for energy production or as a building block for other chemical manufacturing processes; and 2) the ability to accommodate a wide variety of gaseous, liquid, and solid feedstocks. Conventional fuels such as coal and oil, as well as low- or negative-value materials and wastes such as petroleum coke, heavy refinery residuals, secondary oil-bearing refinery materials, municipal sewage sludge, hydrocarbon contaminated soils, and chlorinated hydrocarbon byproducts have all been used successfully in gasification operations.

Gasification of these materials has many potential benefits when compared with conventional options such combustion or disposal by incineration. Recently, the U.S. Environmental Protection Agency (EPA) announced that the Agency is considering an exclusion from the Resource Conservation and Recovery Act (RCRA) for listed secondary oil-bearing refinery materials when processed in a gasification system, an exclusion analogous to the one granted for insertion of RCRA listed refinery wastes into the coking process at refineries. In addition, representatives of the gasification industry have asked EPA to consider a broader exclusion that would include gasification of any carbonaceous material, including hazardous wastes from other industrial sectors (e.g., chemical manufacturing) in modern, high-temperature slagging gasifiers.

The purpose of this report is to provide an independent, third-party description of waste gasification and to present information that clearly defines the differences between the modern gasification and incineration technologies. The primary focus of this document is the currently proposed exemption for gasification of secondary oil-bearing materials in refineries. The objectives of this report are to:

- Compare and contrast the process unit operations and chemical reaction mechanisms of gasification and incineration;

- Cite environmental and regulatory concerns currently applicable to hazardous waste incineration processes and relate them to gasification processes; and

- Provide a summary of existing process stream characterization data for gasification including information on the data quality, sampling/analytical method applicability, and method development needs.

Conclusions

Both gasification and incineration are capable of converting hydrocarbon-based hazardous materials to simple, nonhazardous byproducts. However, the conversion mechanisms and the nature of the byproducts differ considerably, and these factors should justify the separate treatment of these two technologies in the context of environmental protection and economics.

Modern, high temperature slagging gasification technologies offer an alternative process for the recovery and recycling of low-value materials by producing a more valuable commodity, syngas. The multiple uses of syngas (power production, chemicals, methanol, etc.) and the availability of gas cleanup technologies common to the petroleum refining industry make gasification of secondary oil-bearing materials a valuable process in the extraction of products from petroleum. By producing syngas, sulfur, and metal-bearing slag suitable for reclamation, wastes are minimized and the emissions associated with their destruction by incineration are reduced.

Data on syngas composition from the gasification of a wide variety of feedstocks (oil, petroleum coke, coal, and various hazardous waste blends) indicates the major components of syngas to consistently be CO, H_2, and CO_2 with low levels of N_2 and CH_4 also present. Hydrogen sulfide levels in the raw syngas are related to the sulfur content of the feedstock. Similarly, NH_3 and HCN concentrations are related to the fuel's nitrogen content, and HC1 levels are affected by the fuel's chlorine content.

Organic compounds such as benzene, toluene, naphthalene, and acenaphthalene have been detected at very low levels in the syngas from some gasification systems. However, when the syngas is used as a fuel and combusted in a gas turbine, the emissions of these compounds or other organic HAPs are either not detected or present at sub-part-per-billion concentrations in the emitted stack gas. In addition, emissions of particulate matter are found to be one to two orders of magnitude below the current RCRA emissions standards and the recently proposed MACT standard for hazardous waste incinerators.

Although comprehensive test data from the gasification of coal and other fossil fuels are available to assess the fate of many hazardous constituents, the same type and volume of data for the gasification of hazardous wastes are not readily available. To fully assess the performance of gasification on a broader spectrum of hazardous wastes, additional testing may be required to fill data gaps and provide validation of test methods.

All things considered, the ability of gasification technologies to extract useful products from secondary oil-bearing materials and listed refinery wastes is analogous to petroleum coking operations and unlike hazardous waste incineration. Like petroleum coking, gasification can be viewed as an integral part of the refining process where secondary oil-bearing materials can be converted to a syngas that is of comparable quality to the syngas produced from the gasification of fossil fuels.

Table of Contents

List of Tables

List of Figures

Glossary

APCD	Air Pollution Control Device
API	American Petroleum Institute
BIF	Boiler and Industrial Furnace
CAA	Clean Air Act
CWCGP	Cool Water Coal Gasification Program
DRE	Destruction and Removal Efficiency
DAF	Dissolved Air Flotation
EDF	Environmental Defense Fund
EPA	Environmental Protection Agency
ESP	Electrostatic precipitator
ETC	Environmental Technology Council
GTC	Gasification Technologies Council
HAP	Hazardous Air Pollutant
HRSG	Heat Recovery Steam Generator
HWI	Hazardous Waste Incinerator
IGCC	Integrated Gasification Combined Cycle
IWS	Ionizing Wet Scrubber
KDHE	Kansas Department of Health and Environment
LGTI	Louisiana Gasification Technology Inc.
MACT	Maximum Achievable Control Technology
mg/dscm	Milligrams per Dry Standard Cubic Meter
MMscfd	Million Standard Cubic Feet per Day
NESHAP	National Emission Standards for Hazardous Air Pollutants
NODA	Notice Of Data Availability
PAH	Polycyclic Aromatic Hydrocarbons
PCDDs	Polychlorinated Dibenzo(p)dioxins
PCDFs	Polychlorinated Dibenzofurans
PIC	Products of Incomplete Combustion
POHC	Principal Organic Hazardous Constituent
POTW	Publicly Owned Treatment Works
RCRA	Resource Conservation and Recovery Act
SITE	Superfund Innovative Technology Evaluation
SVOCs	Semi-volatile Organic Compounds
SWS	Sour Water Stripper
TCLP	Toxicity Characteristic Leaching Procedure
VOCs	Volatile Organic Compounds
WET–STLC	Waste Extraction Test—Soluble Threshold Limit Concentration
WWT	Waste Water Treatment

Glossary (continued)

Chemical Formulas

Ag	Silver
As	Arsenic
Ba	Barium
Be	Beryllium
Cd	Cadmium
CH_4	Methane
Cl_2	Free chlorine
Co	Cobalt
CO	Carbon monoxide
CO_2	Carbon dioxide
COS	Carbonyl sulfide
Cr	Chromium
Cu	Copper
H_2	Hydrogen
H_2O	Water
H_2S	Hydrogen sulfide
HCl	Hydrogen chloride
HF	Hydrogen fluoride
Hg	Mercury
$HgCl_2$	Mercuric chloride
Mn	Manganese
Mo	Molybdenum
NH_3	Ammonia
Ni	Nickel
NO_x	Oxides of nitrogen
O_2	Oxygen
Pb	Lead
Sb	Antimony
Se	Selenium
SO_2	Sulfur dioxide
SO_3	Sulfur trioxide
SO_x	Oxides of sulfur
Tl	Thallium

Executive Summary

General

Gasification is a technology that has been widely used in commercial applications for more than 50 years in the production of fuels and chemicals. Current trends in the chemical manufacturing and petroleum refinery industries indicate that use of gasification facilities to produce synthesis gas ("syngas") will continue to increase. Attractive features of the technology include: 1) the ability to produce a consistent, high-quality syngas product that can be used for energy production or as a building block for other chemical manufacturing processes; and 2) the ability to accommodate a wide variety of gaseous, liquid, and solid feedstocks. Conventional fuels such as coal and oil, as well as low- or negative-value materials and wastes such as petroleum coke, heavy refinery residuals, secondary oil-bearing refinery materials, municipal sewage sludge, hydrocarbon contaminated soils, and chlorinated hydrocarbon byproducts have all been used successfully in gasification operations.

Gasification of these materials has many potential benefits when compared with conventional options such combustion or disposal by incineration. Recently, the U.S. Environmental Protection Agency (EPA) announced that the Agency is considering an exclusion for the Resource Conservation and Recovery Act (RCRA) for listed secondary oil-bearing refinery materials when processed in a gasification system, an exclusion analogous to the one granted for insertion of RCRA listed refinery wastes into the coking process at refineries. In addition, representatives of the gasification industry have asked EPA to consider a broader exclusion that would include gasification of any carbonaceous material, including hazardous wastes from other industrial sectors (e g , chemical manufacturing) in modern, high-temperature slagging gasifiers. An entrained bed, slurry fed gasifier is the first such unit to process listed refinery wastes without a RCRA Part B permit. The Kansas Department of Health & Environment (KDHE) and EPA agreed in May 1995 that a Part B permit was not required (1).

The purpose of this report is to provide an independent, third-party description of waste gasification and to present information that clearly defines the differences between the modern gasification and incineration technologies. The primary focus of this document is the currently proposed exemption for gasification of secondary oil-bearing materials in refineries. The objectives of this report are to:

- Compare and contrast the process unit operations and chemical reaction mechanisms of gasification and incineration;

- Cite environmental and regulatory concerns currently applicable to hazardous waste incineration process and relate them to gasification processes; and

- Provide a summary of existing process stream characterization data for gasification including information on the data quality, sampling/analytical method applicability, and method development needs.

The EPA has also recently finalized the RCRA Comparable Fuels Exclusion which contains a specific provisions for syngas produced from gasification of hazardous wastes. Under this provision, the syngas is excluded from RCRA requirements if it meets certain specifications for Btu content, total halogen content, total nitrogen content, hydrogen sulfide content, and Appendix VIII trace level constituents. Specific requirements regarding sampling and analysis of the product syngas must meet compliance with the syngas specifications demonstrated before the syngas fuel can be managed as an excluded waste.

Technology Comparison

For the purpose of comparison, the major subsystems used in incineration and gasification technologies can be grouped into four broad categories: 1) Waste preparation and feeding; 2) Combustion vs. gasification; 3) Combustion gas cleanup vs. syngas cleanup; and 4) Residue and ash/slag handling.

Although the major subsystems for incineration and gasification can be grouped in a similar way, the unit operations and fundamental chemical reactions that occur within each major subsystem are very different, perhaps with the exception of waste preparation. Some of the key differences between the two technologies are summarized in Table ES-1.

Four major types of combustion chamber designs are used in modern incineration systems: liquid injection, rotary kiln, fixed hearth, and fluidized bed. Boilers and industrial furnaces (BIF units) are also examples of incineration systems; however, according to EPA

Table ES-1. Key Differences between Gasification and Incineration

Subsystem	Incineration	vs.	Gasification
Combustion vs. Gasification	Designed to maximize the conversion of feedstock to CO_2 and H_2O		Designed to maximize the conversion of feedstock to CO and H_2
	Large quantities of excess air		Limited quantities of oxygen
	Highly oxidizing environment		Reducing environment
	Operated at temperatures below the ash melting point. Mineral matter converted to bottom ash and fly ash.		Operated at temperatures above the ash melting point. Mineral matter converted to glassy slag and fine particulate matter (char).
Gas Cleanup	Flue gas cleanup at atmospheric pressure		Syngas cleanup at high pressure.
	Treated flue gas discharged to atmosphere		Treated syngas used for chemical production and/or power production (with subsequent flue gas discharge).
	Fuel sulfur converted to SO_x and discharged with flue gas.		Recovery of reduced sulfur species in the form of a high purity elemental sulfur or sulfuric acid byproduct.
Residue and Ash/Slag Handling	Bottom ash and fly ash collected, treated, and disposed as hazardous wastes.		Slag is non-leachable, non-hazardous and suitable for use in construction materials. Fine particulate matter recycled to gasifier or processed for metals reclamation.

MACT information, less than 15% of the hazardous waste is disposed of in these units. The application of each type of combustion chamber is a function of the physical form and ash content of the wastes being combusted. In each of these designs, waste material is combusted in the presence of a relatively large excess of oxygen (air) to maximize the conversion of the hydrocarbon-based wastes to carbon dioxide and water (50% to 200%). In some configurations, excess fuel and oxygen must be added to increase incineration temperatures to improve destruction and removal efficiency. This also increases the production and emission of carbon dioxide.

Sulfur and nitrogen in the feedstock are oxidized to form SO_x and NO_x. Halogens in the feedstock are primarily converted to acid halide gases such as HCl and HF and exit the combustion chamber with the combustion gases. Temperatures in the refractory-lined combustion chambers may range from 1200°F to 2500°F with mean gas residence times of 0.3 to 5.0 seconds (2,3).

Incinerators typically operate at atmospheric pressure and temperatures at which the mineral matter or ash in the waste is not completely fused (as slag) during the incineration processes. Ash solids will either exit the bottom/discharge end of the combustion chambers as bottom ash or as particulate matter entrained in the combustion flue gas stream.

Combustion gases from hazardous waste incineration systems are typically processed in a series of treatment operations to remove entrained particulate matter, heavy metals, and acid gases such as HCl and other inorganic acid halides. Systems that process low ash or low halogen content liquid wastes may not require any downstream process controls. However, one of the more common gas cleanup configurations used at waste incineration facilities is a gas quench (gas cooling), followed by a venturi scrubber (particulate removal) and a packed tower absorber (acid gas removal). Wet electrostatic precipitators and ionizing wet scrubbers are used at some facilities for combined particulate and acid gas removal. Fabric filter systems are also used for particulate removal in some applications. Demisters are often used to treat the combustion gases before they are discharged to the atmosphere to reduce the visible vapor plume at the stack. These cleanup systems typically operate at atmospheric pressure and must process a large volume of flue gas produced as a result of the large excess air requirements of incineration systems.

The GTC, in response to comments received by EPA on the Notice of Data Availability regarding the proposed refinery gasification exclusion (63 FR 38139, July 15, 1998), has proposed the following definition of "gasification" for the purpose of qualifying for this exclusion:

- A process technology that is designed and operated for the purpose of producing synthesis gas (a commodity which can be used to produce fuels, chemicals, intermediate products, or power) through the chemical conversion of carbonaceous materials.

- A process that converts carbonaceous materials through a process involving partial oxidation of the feedstock in a reducing atmosphere in the presence of steam at temperatures sufficient to convert the feedstock to synthesis gas, to convert inorganic matter in the feedstock (when the feedstock is a solid or semi-solid) to a glassy solid material known as vitreous frit or slag, and to convert halogens into the corresponding acid halides.

- A process that incorporates a modern, high-temperature pressurized gasifier (which produces a raw synthesis gas) with auxiliary gas and water treatment systems to produce a refined product synthesis gas, which when combusted, produces emissions in full compliance with the Clean Air Act.

Modern gasification systems that meet the GTC definition of gasification as presented above, are applicable to refinery and chemical manufacturing operations, as well as IGCC power systems. These gasification systems can be categorized as either entrained bed or moving/fixed bed. The gasification process described by this definition operates by feeding carbon-containing materials into a heated and pressurized chamber (the gasifier) along with a controlled and limited amount of oxygen and steam. At the high operating temperature and pressure created by conditions in the gasifier, chemical bonds are broken by oxidation and steam reforming at temperatures sufficiently high to promote very rapid reactions. Inorganic mineral matter is fused or vitrified to form a molten glass-like substance called slag or vitreous frit. With insufficient oxygen, oxidation is limited and the thermodynamics and chemical equilibria of the system shift reactions and vapor species to a reduced, rather than an oxidized state. Consequently, the elements commonly found in fuels and other organic materials (C, H, N, O, S, Cl) end up in the syngas as the following compounds: CO, H_2, H_2O, CO_2, N_2, CH_4, H_2S, and HCl with lesser amounts of COS, NH_3, HCN, elemental carbon, and trace quantities of other hydrocarbons. The reducing atmosphere within the gasification reactor prevents the formation of oxidized species such as SO_2 and NO_x.

A wide variety of carbonaceous feedstocks can be used in the gasification process including: coal, heavy oil, petroleum coke, orimulsion, and waste materials (e.g., refinery wastes, contaminated soils, chlorinated wastes, municipal sewage sludge, etc.). Low-Btu wastes may be blended with high-Btu content supplementary fuels such as coal or petroleum coke to maintain the desired gasification temperatures in the reactor. However, unlike incineration, these supplementary fuels contribute primarily to the production of more syngas and not to the production of CO_2.

After the gasification step, the raw synthesis gas temperature is reduced by quenching with water, slurry, and/or cool recycled syngas. Further cooling may be done by heat exchange in a syngas cooler before entrained particulate is removed. Particulate matter is captured in the water and filtered from the water if direct-water scrubbing is utilized. Alternatively, particulates may be removed via dry filtration or hot gas filtration. Moisture in the syngas condenses as it is cooled below its dewpoint. Any particulate scrubber water and syngas cooling condensates contain some water-soluble gases (NH_3, HCN, HCl, H_2S). Further refinement of the syngas is conditional upon the end use of the product syngas but usually includes the removal of sulfur compounds (H_2S and COS) for the recovery of high-purity sulfur as a marketable product. Sulfur removal and recovery are accomplished using commercially available technologies common to the refinery and natural gas industries.

Byproduct Utilization and Treatment

Gasification and incineration technologies are significantly different in terms of byproduct utilization and treatment. Table ES-2 provides a summary of the byproduct and emission streams for each technology.

Slag is the primary solid byproduct of gasification and the quantity produced is a function of how much mineral matter is present in the gasifier feeds. The slag contains mineral matter

Table ES-2. Comparison of Byproduct and Emission Sources for Gasification and Incineration Processes

Process Subsystem	Gaseous		Liquid		Solid	
	Gasification	Incineration	Gasification	Incineration	Gasification	Incineration
Waste/Fuel Preparation					• Fuel/waste rejects	• Fuel/waste rejects
Combustor vs. Gasifier	• **Steam***	• **Steam***			• **Slag***	• Bottom ash
Gas Cleanup	• **Clean synthesis gas***	• Combustion stack gas	• **High purity sulfur***		• Spent sulfur recovery catalysts • Solvent filter cake residues	
Residue and Slag/Ash Handling	• Tail-gas incinerator stack from sulfur recovery system		• Treated process water	• Treated process water	• Fine particulate matter* • WWT sludge	• Fly ash • WWT sludge
End Use Processes (e.g., IGCC power production)	• Combustion turbine/HRSG stack gas • **Steam*** • **Electricity***					

* Bold type indicates a byproduct stream which can be sold, used as feedstock in downstream chemical production processes, or recycled in other in-plant process operations.

associated with the feed in a vitrified form, a hard glassy substance. This is the result of gasifier operation at temperatures above the fusion or melting temperature of the mineral matter. Thus, feeds such as coal produce much more slag than petroleum feedstocks (heavy oil, petroleum coke, etc.). Because the slag is in a fused, vitrified state, it rarely fails the TCLP for metals. Slag is not a good substrate for binding organic compounds so it is usually found to be nonhazardous, exhibiting none of the characteristics of hazardous waste. Thus, it may be disposed of in a landfill or sold as an ore to recover the metals concentrated within its structure. Slag's hardness also makes it suitable as an abrasive or additive in road-bed construction materials.

Downstream of the gasifier, unconverted fines and light-ash material are removed from the raw syngas using wet scrubbers or dry filtration processes. The fine particulate matter often contains a high percentage of carbon, so the material is often recycled to the gasifier to recover the energy value of this material. In the case of refinery applications, the petroleum feedstocks can contain high levels of nickel and vanadium. These elements are concentrated in the fine particulate matter exiting the gasifier with the raw syngas. Thus, the fine particulate matter removed from the syngas is processed further to recover these metals. A number of metals recovery processes are currently in use and typically involve separation of the solids from the scrubber water (if wet removal techniques are used), drying of the solids, and controlled combustion of the solids in a furnace to oxidize vanadium compounds to vanadium pentoxide, a product that can be sold for use in the metalurgical industry. The resulting product may contain up to 75 weight percent vanadium, depending on the composition of the feed materials (4,5,6,7).

Sulfur compounds (H_2S and COS) in the particulate-free syngas are typically removed and recovered using conventional gas treatment technologies from the refinery and natural gas industries. The resulting byproduct is high-purity liquid sulfur. Sulfur removal efficiencies on the order of 95 to 99% are typically achieved using these systems. The clean product syngas can then be used as fuel to a combustion turbine to produce electricity, processed as a source of hydrogen, and/or used as a feedstock for the production of other chemical products. The portion of the clean syngas combusted in a gas turbine is the major source of gaseous emission for the process.

The various water streams resulting from syngas cooling and cleaning are typically recycled to the gasifier or to the scrubber after entrained solids have been removed. A small portion of the water must be purged from the system to avoid accumulation of dissolved salts. One commonly used method for treatment of this process water offers an additional opportunity to recover sulfur that is present in the water in the form of dissolved gases. The process water is "flashed" in a vessel at low pressure to release the dissolved gases, and the flash gas is route to the sulfur removal unit with the raw syngas.

The resulting water is then recycled to the process or a portion blown down to a conventional waste water treatment system. Gas condensate may also be steam-stripped to remove ammonia, carbon dioxide, and hydrogen sulfide. Stripped water is recycled to the process. The resulting stripper overhead gas may be routed to the sulfur recovery unit or incinerated along with the tail gas from the sulfur recovery unit. Flue gases from the tail-gas incinerator are released to the atmosphere subject to permit limitations for such things as SO_2 and NO_x.

Environmental Characterization Data

A. SO_x, NO_x and Particulate Matter

For a given secondary material, emission levels of SO_x, NO_x, and particulate from gasification systems are reduced significantly compared to incineration systems. In an oxidative incineration environment, sulfur and nitrogen compounds in the feed are converted to SO_x and NO_x. In contrast, syngas cleanup systems for modern gasification systems are designed to recover 95 to 99% of the sulfur in the fuel as a high-purity sulfur byproduct. Likewise, nitrogen in the feed is converted to diatomic nitrogen (N_2) and ammonia in the syngas. Ammonia is subsequently removed from the syngas in downstream cleanup systems such as particulate scrubbing and gas cooling. Thus, if the clean syngas is combusted in a gas turbine to generate electricity, the production of SO_x and NO_x is reduced significantly. If the syngas is used as feedstock in downstream chemical manufacturing processes, these compounds are not formed. Data for repowering of coal-fired electric utilities with IGCC technology has shown that emissions of SO_x, NO_x, and particulate are reduced by one to two orders of magnitude (8).

Typical end uses for the clean syngas from gasification systems (e.g., electricity production in a gas turbine or chemical manufacturing feedstock) require a product syngas with very low particulate content. Particulate levels in the raw syngas are reduced to very low levels because of the multiple gas cleanup systems used in gasification systems. Particulate scrubbers or dry filtration systems are used for primary removal of particulate matter. Often, this captured particulate matter is recycled to the gasifier.

Additional particulate removal occurs in the gas cooling operations and in the acid gas removal systems used to condition and recover sulfur from the raw syngas. As a result, measured particulate emissions at coal-fired gasification systems where the clean syngas was combusted in a turbine are two orders of magnitude lower than the existing RCRA standard for hazardous waste incinerators (RCRA limit = 180 mg/dcsm), and one order of magnitude below the recently finalized MACT limit for new and existing hazardous waste incinerators (MACT limit = 34 mg/dscm) (9,10,11). Particulate matter concentrations less than 10 mg/dscm in the gas turbine emissions have been reported for a gasification system using heavy refinery residual feedstocks such as vacuum visbroken residue, vacuum residue, and asphalt (12).

B. Organic Compounds

Historically, organic compound emissions of most concern from waste incineration systems have been principal organic hazardous constituent (POHC) in the waste feed and products of incomplete combustion (PICs). Air emissions of these compounds have been characterized extensively for hazardous waste incinerators. POHC refers to the organic compounds present in the waste feed that must be destroyed at greater than 99.99% efficiency (99.9999% for listed dioxin wastes) based on RCRA rules for hazardous waste incineration systems. PICs are compounds such as semi-volatile organic compounds (SVOCs), polycyclic aromatic hydrocarbons (PAHs), VOCs, and dioxin/furan compounds.

EPA's database for hazardous waste incinerators includes data for 46 SVOCs and 59 VOCs detected in the combustion gases over a wide range of concentration (13). The VOCs tend to be detected more often and at higher concentrations than the SVOCs. Dioxin/furan compounds (PCDDs/PCDFs) are also often detected in the combustion gases from hazardous waste incinerators. Therefore, specific concentration-based limits for these compounds have been established in the recently finalized MACT rules for hazardous waste incinerators (9).

Similar data for gasifier product syngas and turbine/HRSG stack emissions are much more limited. The most comprehensive trace substance characterization tests have been conducted for entrained bed and two-stage entrained bed gasifiers using both slurry and dry feed systems (10,11,14,15). These studies were conducted during the gasification of various coal feedstocks and did not include gasification of secondary materials. Less comprehensive test data are also available for refinery gasification operations (12,16,17,18) and waste gasification processes (19,20,21,22,23).

One of the most applicable data sets can be found in a Technology Evaluation Report prepared in 1995 by Foster-Wheeler Enviresponse, Inc. (FWEI) under the EPA Superfund

Innovative Technology Evaluation (SITE) Program (24). The report presents an evaluation of a slurry fed, single stage, entrained bed gasifier feeding a coal-soil-water fuel with chlorobenzene added as a POHC to measure the destruction and removal efficiency (DRE) of the process. Lead and barium salts were also added to track the fate of these and other heavy metals. The report from the SITE program also briefly describes the results of additional gasification tests using secondary materials such as refinery tanks bottoms, municipal sewage sludge, and hydrocarbon-contaminated soils.

Results from these measurement programs are summarized in Table ES-3. In general, VOCs such as benzene, toluene, and xylene, when detected, were present a parts per billion levels. SVOCs, including PAHs, were also detected in the sygas and/or turbine exhaust/tail gas incinerator stack in some cases. SVOCs were typically present at extremely low levels on the order of parts per trillion.

Gasification tests using chlorinated feedstocks have also been conducted to measure the DRE for organic compounds such as chlorobenzene and hexachlorobenzene (20,24). Destruction and removal efficiencies greater than 99.99% were demonstrated for both compounds for an entrained bed and a fixed bed gasifier.

Dioxin and furan compounds (PCDD/PCDFs) are not expected to be present in the syngas from gasification systems for two reasons. First, the high temperatures in the gasification process effectively destroy any PCDD/PCDF compounds or precursors in the feed. Secondly, the lack of oxygen in the reduced gas environment would preclude the formation of the free chlorine from HCl, thus limiting chlorination of any precursors in the syngas. Measurements of PCDD/PCDF compounds in gasification systems confirm these expectations as shown in Figure ES-1. The configuration of the gasification systems represented in Figure ES-1 are as follows:

Site A – EPA SITE program. Gasification of RCRA soil/coal mixture including chlorobenzene. Entrained bed gasifier.

Site B – Fixed bed waste gasifier.

Site C – Waste gasification facility in Germany. Fixed and entrained bed gasifiers.

Table ES-3. Organic Compound Measurements for Various Gasification Processes

Test Program	System Configuration	Fuel Type	Syngas	Turbine Exhaust and/or Tail Gas Incinerator Stack
CWCGP (10)	Entrained bed, slurry feed, wet scrubber, Selexol, SCOT/Claus	Illinois 6, SUFCO, Lemington, and Pitt. 8 coals	NR	PAHs, SVOCs not detected. Benzene, toluene, occasionally detected at ppbv levels.
LGTI (11)	Two-stage entrained bed, slurry feed, wet scrubber, Selectamine™, Seletox™/Claus	Powder River Basin coal	NR	Benzene, toluene detected at sub-ppbv. PAHs detected at pptv.
SCGP-1 (14,15)	Entrained bed, dry feed, dry particulate collection, wet scrubber, Sulfinol™, SCOT/Claus	Illinois 5, Blacksville, Drayton, El Cerrejon coals	PAHs and phenolics not detected (DL ~ 1 ppbv). Total other non-methane hydrocarbons detected at 0.5 to 90 ppbw in raw syngas.	NR
SITE (24)	Entrained bed, slurry feed, wet scrubber, Selexol, sodium hydroxide acid gas absorber, pilot-scale	Chlorobenzene RCRA soil/Pitt. 8 coal	Selected VOCs and PAHs detected at sub-ppbv concentrations in raw and clean syngas. 99.9956% DRE	NR
Other (24)	Entrained bed, slurry feed, wet scrubber, Selexol, sodium hydroxide acid gas absorber, pilot-scale	Refinery tank bottoms/coal, MSW/coal, Hydrocarbon soils/coal	No organic compounds heavier than methane detected at > 1 ppmv.	NR
RCl (22)	Entrained bed, HCl byproduct recovery	100% Chlorinated heavies DCP and DCE	Chlorinated VOCs not detected (DL ~ 1 ppbv). Benzene, toluene, ethylbenzene and xylenes detected at ppbv levels.	NR
SGI (20)	Fixed bed, dry feed, pilot-scale	Hexachlorobenzene and petroleum coke	99.9999% DRE	NR

NR = Not reported.

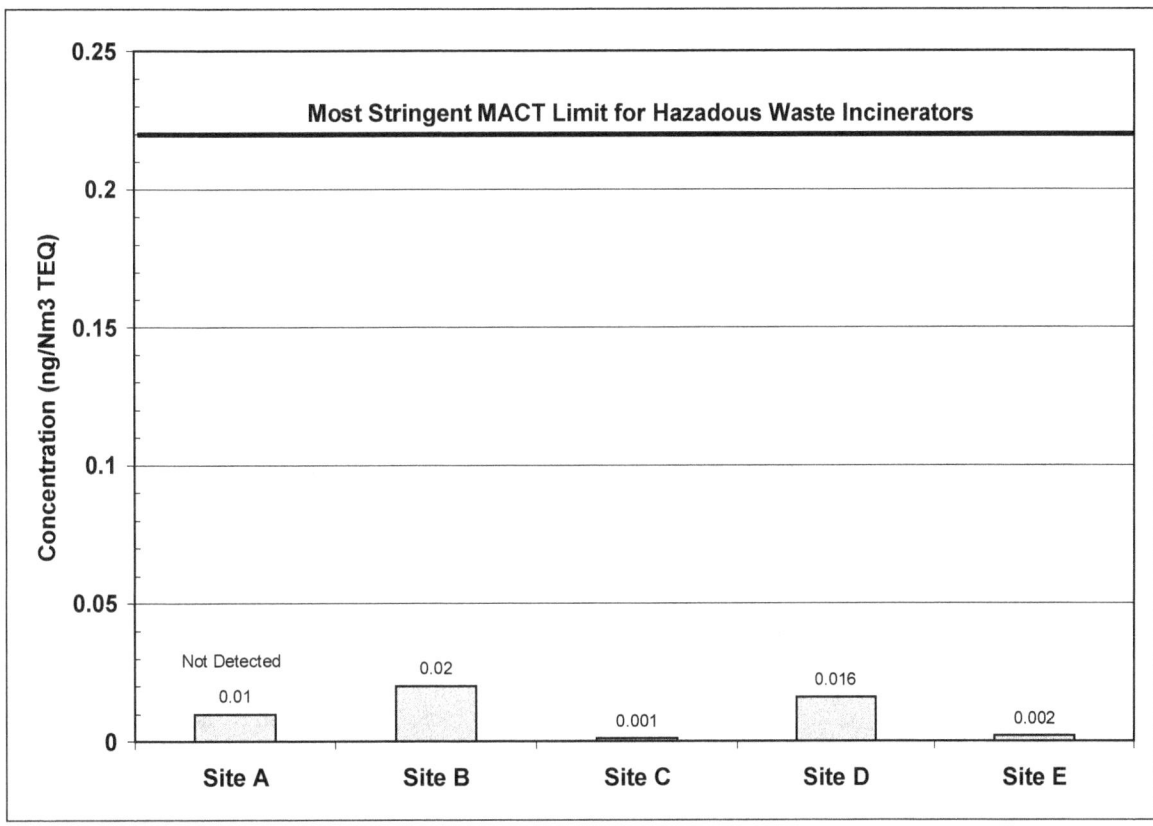

Figure ES-1. Measured Concentrations of PCDD/PCDF Compounds in Syngas Produced from Gasification

Site D – RCl process for gasification of 100% chlorinated heavies from manufacture of DCP and DCE. Entrained bed gasifier.

Site E – Demonstration of PCB destruction in a fixed bed gasifier. Hexachlorobenzene and petroleum coke feeds.

In all cases, the levels of PCDD/PCDF compounds were one to two orders of magnitude below the most stringent MACT standard recently finalized for hazardous waste incinerators (0.22 ng/Nm3 TEQ).

C.   Trace Metals and Halides

Gas Streams. EPA data for hazardous waste incineration systems indicate that metals emissions include antimony, arsenic, beryllium, cadmium, chromium, lead, mercury, nickel, and selenium compounds (13,26). Acid halides (HCl, HF and HBr) may also be present depending on the halogen content of the waste feed. Specific concentration-based emission limits have been established for specific trace metals or groups of metals in the recently finalized MACT rules for hazardous waste incinerators (9).

Review of the available literature shows that a comprehensive characterization of trace elements has not been conducted for gasification technologies feeding secondary materials. Thus, specific conclusions regarding the level of trace constituents in the syngas, or those emitted from gas turbine stack and tail-gas incinerator stacks during gasification of secondary materials, cannot be directly drawn. However, the data from comprehensive test programs at coal-fired, entrained bed (10,11,14,15) and the EPA SITE program tests do provide valuable insight on the general fate of toxic substances in gasification systems, particularly for metals. A substantial amount of information was collected regarding the partitioning of selected volatile/semi-volatile and non-volatile elements among the various discharge streams.

Based on review of these data, certain trace metals have the potential to be present in the clean syngas or gas turbine exhaust. These metals include: chloride, fluoride, mercury, arsenic, cadmium, lead, chromium, nickel, and selenium. In most cases, the amount of these elements present in the syngas or combustion turbine exhaust represented less than 10% of the amount input to the gasifier with the coal. Elements such as chloride and fluoride are typically removed in the gas scrubbing and cooling operations and ultimately partition primarily to the process water streams. Greater than 99% removal of HCl was measured during the SITE test program. Semi-volatile metals such as lead will tend to volatilize in the gasifier and recondense on the fine particulate matter which is removed from the syngas, resulting in enrichment of these elements.

Mass balance closures for the volatile and semi-volatile trace elements tend to be substantially less than 100% for all test programs. Thus, the fate of these substances is less certain. However, in one instance, the low recoveries were shown to be evidence of retention of volatile trace elements within the process equipment deposits. There is also evidence to suggest that some of the volatile elements may accumulate in the solvents used in the sulfur removal systems at gasification facilities.

Non-volatile elements such as barium, beryllium, chromium, cobalt, manganese, nickel, and vanadium partition almost entirely to the slag where they are immobilized in the vitrified matrix.

Solids. For hazardous waste incinerators, RCRA requirements mandate that any ash from combustion chamber and downstream gas cleanup devices is also considered a hazardous waste. The principal contaminants are heavy metals primarily in the form of metal oxides and undestroyed organic material. Leaching of heavy metals from incinerator ash material is of particular concern. Test data suggest that very small amounts of residual organic compounds remain in incinerator ash and control device residuals. When organic compounds were detected, they tended to be toluene, phenol, and naphthalene at concentrations less than 30 parts per billion (27,28).

Analysis of the slag material produced from various gasification processes has consistently shown the slag to be a nonhazardous waste according to RCRA definitions. Non-volatile trace metals tend to concentrate in the slag; however, the glassy slag matrix effectively immobilizes the metals eliminating or reducing their leachability. For example, the slag and fine particulate matter produced from the gasification of secondary refinery materials at the El Dorado refinery did not exhibited any of the RCRA waste characteristics and were classified as nonhazardous (16). Data from the SITE program and other gasification tests using mixtures of coal and secondary materials (i.e., petroleum tank bottoms, municipal sewage sludge, and hydrocarbon contaminated soils) have shown similar results for the slag. Tests conducted on the fine particulate matter removed from the raw syngas during these test programs indicate that this low-volume material has the potential to exceed TCLP limits for some metals. However, the high carbon and metals content of this material make it a valuable byproduct that is often recycled to the gasifier to recover the energy content or processed to reclaim metals, such as nickel and vanadium when heavy refinery feedstocks are gasified.

Conclusions

Both gasification and incineration are capable of converting hydrocarbon-based hazardous materials to simple, nonhazardous byproducts. However, the conversion mechanisms and the nature of the byproducts differ considerably, and these factors should justify the separate treatment of these two technologies in the context of environmental protection and economics.

Gasification technologies meeting the definition proposed by the GTC offer an alternative process for the recovery and recycling of low-value materials by producing a more valuable commodity – syngas. The multiple uses of syngas (power production, chemicals, methanol, etc.) and the availability of gas cleanup technologies common to the petroleum refining industry make gasification of secondary oil-bearing materials a valuable process in the extraction of products from petroleum. By producing syngas, sulfur, and metal-bearing slag suitable for reclamation, wastes are minimized and the emissions associated with their destruction by incineration are reduced.

Data on syngas composition from the gasification of a wide variety of feedstocks (oil, petroleum coke, coal, and various hazardous waste blends) indicates the major components of syngas to consistently be CO, H_2, and CO_2 with low levels of N_2 and CH_4 also present. Hydrogen sulfide levels in the raw syngas are related to the sulfur content of the feedstock. Similarly, NH_3 and HCN concentrations are related to the fuel's nitrogen content, and HCl levels are affected by the fuel's chlorine content.

Organic compounds such as benzene, toluene, naphthalene, and acenaphthalene have been detected at very low levels in the syngas from some gasification systems. However, when used as a fuel and combusted in a gas turbine, the emissions of these compounds or other organic HAPs are either not detected or present at sub-part-per-billion concentrations in the emitted stack gas. In addition, emissions of particulate matter are found to be one to two orders of magnitude below the current RCRA emissions standards and the recently proposed MACT standard for hazardous waste incinerators.

Although comprehensive test data from the gasification of coal and other fossil fuels are available to assess the fate of many hazardous constituents, the same type and volume of data for the gasification of hazardous wastes are not readily available. To fully assess the performance of gasification on a broader spectrum of hazardous wastes, additional testing may be required to fill data gaps and provide validation of test methods.

All things considered, the ability of gasification technologies to extract useful products from secondary oil-bearing materials and listed refinery wastes is analogous to petroleum coking operations and unlike hazardous waste incineration. Like petroleum coking, gasification can be viewed as an integral part of the refining process where secondary oil-bearing materials can be converted to a fuel (syngas) that is of comparable quality to the syngas produced from the gasification of fossil fuels.

References

1. Rhodes, A.K. "Kansas Refinery Starts Up Coke Gasification Unit," *Oil & Gas Journal*, August 5, 1996.

2. Oppelt, T.E. "Incineration of Hazardous Waste: A Critical Review," *JAPCA*, Vol. 37, No. 5, May 1987.

3. Dempsey, C. and , T.E, Oppelt. "Incineration of Hazardous Waste: A Critical Review Update," *Air and Waste*, Vol. 43, January 1993.

4. Liebner, W., "MGP-Lurgi/SVZ Mulit Purpose Gasification, Another Commercially Proven Gasification Technology," Presented at the 1999 Gasification Technologies Conference, San Francisco, CA, October 17-20, 1999.

5. De Graaf, J.D., E.W. Koopmann, and P.L. Zuideveld, "Shell Pernis Netherlands Refinery Residue Gasification Project," Presented at the 1999 Gasification Technologies Conference, San Francisco, CA, October 17-20, 1999.

6. The Shell Gasification Process. Vendor literature process description. Shell Global Solutions U.S. Houston, TX, October 1999.

7. Maule, K. and S. Kohnke, "The Solution to the Soot Problem in an HVG Gasification Plant," Presented at the 1999 Gasification Technologies Conference, San Francisco, CA, October 17-20, 1999.

8. U.S. Department of Energy. *The Wabash River Coal Gasification Repowering Project*, Topical Report Number 7, November 1996.

9. U.S. EPA. *Final MACT Rule for Hazardous Waste Combustors*, 64 FR 52828, September 30, 1999.

10. Electric Power Research Institute. *Cool Water Coal Gasification Program: Final Report,"* prepared by Radian Corporation and Cool Water Coal Gasification Program. EPRI Final Report GS-6806, December 1990.

11. Electric Power Research Institute. *Summary Report: Trace Substance Emissions from a Coal-Fired Gasification Plant*, prepared for EPRI and the U.S. Department of Energy, June 29, 1998.

12. Collodi, G. and R.M. Jones, *The Sarlux IGCC Project and Outline of the Construction and Commissioning Activities*, Presented at the 1999 Gasification Technologies Conference, San Francisco, CA, October 17-20, 1999.

13. U.S. EPA. *Draft Technical Support Document for HWC MACT Standards Volume II: HWC Emissions Database*, Office of Solids Waste and Emergency Response, February 1996.

14. Baker, D.C., W.V Bush, and K.R. Loos. "Determination of the Level of Hazardous Air Pollutants and other Trace Constituents in the Syngas from the Shell Coal Gasification Process," *Managing Hazardous Air Pollutants: State of the Art*, W. Chow and K.K Conner (eds.), Lewis Publishing, EPRI TR-101890, 1993.

15. Baker, D.C. "Projected Emissions of Hazardous Air Pollutants from a Shell Gasification Process-Combined-Cycle Power Plant," *Fuel,* Volume 73, No. 7, July 1994.

16. DelGrego, G., *Experience with Low Value Feed Gasification at the El Dorado, Kansas Refinery*, Presented at the 1999 Gasification Technologies Conference, San Francisco, CA, October 17-20, 1999.

17. Liebner, W., *MGP-Lurgi/SVZ Mulit Purpose Gasification, Another Commercially Proven Gasification Technology*, Presented at the 1999 Gasification Technologies Conference, San Francisco, CA, October 17-20, 1999.

18. De Graaf, J.D., E.W. Koopmann, and P.L. Zuideveld, "Shell Pernis Netherlands Refinery Residue Gasification Project," Presented at the 1999 Gasification Technologies Conference, San Francisco, CA, October 17-20, 1999.

19. Skinner, F.D., *Comparison of Global Energy Slagging Gasification Process for Waste Utilization with Conventional Incineration Technologies.* Final Report, Radian Corporation, January 1990.

20. Vick, S.C., *Slagging Gasification Injection Technology for Industrial Waste Elimination*, Presented at the 1996 Gasification Technologies Conference, San Francisco, CA, October 1996.

21. Seifert, W., "Utilization of Wastes – Raw Materials for Chemistry and Energy. A Short Description of the SVZ-Technology," Prepared for the technical conference: "Gasification the Gateway to a Cleaner Future" Dresden, Germany. September 23-24, 1998.

22. Salinas, L., P. Bork and E. Timm, *Gasification of Chlorinated Feeds*, Presented at the 1999 Gasification Technologies Conference, San Francisco, CA, October 17-20, 1999.

23. The Thermoselect Solid Waste Treatment Process. Vendor literature supplied by Thermoselect Incorporated. Troy, Michigan, 1999.

24. U.S. EPA. *Texaco Gasification Process Innovative Technology Evaluation Report*, Office of Research and Development Superfund Innovative Technology Evaluation Program, EPA/540/R-94/514, July 1995.

25. Gasification Technology Counsel. Response to Comments in ETC letter of October 13, 1998 and EDF letter of October 13, 1998. Letter for RCRA Docket Number F-98-PR2A-FFFFF, May 13, 1998.

26. U.S. EPA. *Proposed MACT Rule for Hazardous Waste Combustors*, 61 FR 17357, April 19, 1996.

27. U.S. EPA. *Performance Evaluation of Full-Scale Hazardous Waste Incineration*, five volumes, NTIS, PB- 85-129500, November 1994.

28. Van Buren, D., G. Pie, and C. Castaldini, *Characterization of Hazardous Waste Incineration Residuals*, U.S. EPA, January 1987.

1.0 Introduction

Gasification is a technology that has been widely used in commercial applications for over 40 years in the production of fuels and chemicals. Current trends in the chemical manufacturing and petroleum refinery industries indicate that use of gasification facilities to produce synthesis gas ("syngas") will continue to increase. Attractive features of the technology include: 1) the ability to produce a consistent, high quality syngas product that can be used for energy production or as a building block for other chemical manufacturing processes; and 2) the ability to accommodate a wide variety of gaseous, liquid, and solid feedstocks. Conventional fuels such as coal and oil, as well as low-value materials and wastes such as petroleum coke, secondary oil-bearing refinery materials, heavy refinery residues, municipal sewage sludge, hydrocarbon contaminated soils, and chlorinated hydrocarbon by-products have all been used successfully in gasification operations.

The U.S. Department of Energy (DOE) has promoted the continued development of gasification technology because of the superior energy efficiency and environmental performance of the process for energy production applications. Specifically, DOE has focused its efforts on the Integrated Gasification Combined Cycle (IGCC) systems which replace the traditional coal combustor with a gasifier and gas turbine. Exhaust heat from the gas turbine is used to produce steam for a conventional steam turbine, thus the gas turbine and steam turbine operate in a combined cycle. The IGCC configuration provides high system efficiencies and ultra-low pollution levels. SO_2 and NO_x emissions less than one-tenth of that allowed by New Source Performance Standards limits have been demonstrated. DOE has also been involved in the evaluation and development of sampling and analytical methods for the measurement of trace level substances in gasification process streams (e.g., mercury in syngas).

In July of 1998, the U.S. EPA issued a Notice of Data Availability (NODA) announcing that the Agency is considering a RCRA exclusion for gasification of oil-bearing secondary materials in refinery operations (63 FR 38139). Specifically, EPA is assessing whether oil-bearing hazardous secondary materials generated within the petroleum industry should be excluded from the definition of solid waste when inserted into gasification units. The proposed gasification exclusion would be analogous to the RCRA exclusion granted for the insertion of similar refinery secondary materials into the coker process at petroleum refineries (63 FR 42109). The gasification exclusion would apply to any oil-bearing secondary material, including RCRA listed hazardous refinery wastes K048-K052, F037, and F038 (e.g., DAF float, slop oil emulsion solids, heat exchanger cleaning sludge, API separator sludge, tank bottoms, oil/water separation sludge, etc.). In addition, representatives of the gasification industry have asked EPA to a consider a broader exclusion for gasification facilities that would include gasification of any carbonaceous material, including hazardous wastes from other industrial sectors (e.g., chemical manufacturing), in a modern, high temperature slagging gasifier.

Subsequent comments from the Environmental Technology Council (ETC), which represents the hazardous waste incineration industry, and from the Environmental Defense Fund (EDF) regarding the July 1998 NODA revealed a lack of understanding of modern gasification systems. The EPA staff considering the gasification exclusion have also expressed the desire to have information that clearly defines the differences between gasification and incineration of hazardous waste to assist them in their rule making process.

This document has been prepared for the DOE in response to these needs. The purpose of this paper is to provide an independent, third-party description of waste gasification, and to provide DOE and EPA with information that clearly defines the differences between the modern gasification and incineration technologies. The primary focus of this document is the currently proposed exemption for gasification of secondary oil-bearing materials in refineries. The objectives of this report are to:

- Compare and contrast the process unit operations and chemical reaction mechanisms of gasification and incineration;

- Cite environmental and regulatory concerns currently applicable to hazardous waste incineration process and relate them to gasification processes; and

- Provide a summary of existing process stream characterization data for gasification including information on the data quality, sampling/analytical method applicability, and method development needs.

Section 2 provides detailed process descriptions for the major unit operations used in modern gasification and hazardous waste incineration systems. Information regarding specific byproduct and emission streams from gasification and incineration processes, and their possible utilization or treatment is provided in Section 3. A discussion of the auxiliary systems designed to recover or treat the byproducts from both technologies is included. Section 4 identifies the current environmental regulations affecting the incineration of hazardous wastes and any proposed regulations applicable to waste gasification. Finally, Section 5 contains a discussion of the currently available environmental characterization data that exists for gasification systems. Data gaps and method development needs for gasification systems are also identified.

2.0   Process Descriptions

The GTC, in response to comments received by EPA on the Notice of Data Availability regarding the proposed refinery gasification exclusion (63 FR 38139, July 15, 1998), has proposed the following definition of "gasification" for the purpose of qualifying for this exclusion:

- A process technology that is designed and operated for the purpose of producing synthesis gas (a commodity which can be used to produce fuels, chemicals, intermediate products or power) through the chemical conversion of carbonaceous materials.

- A process that converts carbonaceous materials through a process involving partial oxidation of the feedstock in a reducing atmosphere in the presence of steam at temperatures sufficient to convert the feedstock to synthesis gas; to convert inorganic matter in the feedstock (when the feedstock is a solid or semi-solid) to a glassy solid material known as vitreous frit or slag; and to convert halogens into the corresponding acid halides.

- A process that incorporates a modern, high temperature pressurized gasifier (which produces a raw synthesis gas) with auxiliary gas and water treatment systems to produce a refined product synthesis gas, which when combusted, produces emissions in full compliance with the Clean Air Act.

The gasification process described by this definition operates by feeding carbon-containing materials into a heated and pressurized chamber (the gasifier) along with a controlled and limited amount of oxygen and steam. At the high operating temperature and pressure created by conditions in the gasifier, chemical bonds are broken by thermal energy and not by oxidation, and inorganic mineral matter is fused or vitrified to form a molten glass-like substance called slag or vitreous frit. With insufficient oxygen, oxidation is limited and the thermodynamics and chemical equilibria of the system shift reactions and vapor species to a reduced, rather than an oxidized state. Consequently, the elements commonly found in fuels and other organic materials (C, H, N, O, S, Cl) end up in the syngas as the following compounds: CO, H_2, H_2O, CO_2, N_2, CH_4, H_2S, and HCl with lesser amounts of COS, NH_3, HCN, elemental carbon and trace quantities of other hydrocarbons.

After the gasification step, the raw synthesis gas temperature is reduced by quenching with water, slurry and/or cool recycled syngas. Further cooling may be done by heat exchange in a syngas cooler before entrained particulate is removed. Particulate matter is captured in the water and filtered from the water if direct water scrubbing is utilized. Alternatively, particulates may be removed via hot gas dry filtration techniques. Moisture in the syngas condenses as it is cooled below its dewpoint. Any particulate scrubber water and syngas cooling condensates contain some water-soluble gases (NH_3, HCN, HCl, H_2S). Further refinement of the syngas is conditional upon the end use of the product syngas, but usually includes the removal of sulfur compounds (H_2S and COS) for the recovery of sulfur as a marketable product.

Basic block flow diagrams for waste incineration and waste gasification processes are provided in Figures 2-1 and 2-2, respectively, to compare and contrast the two technologies. For the purpose of comparison, the major subsystems used in incineration and gasification have been grouped into four broad categories:

- Waste preparation and feeding;

- Combustion vs. Gasification;

- Combustion Gas Cleanup vs. Syngas Cleanup; and

- Residue and Ash/Slag Handling.

Although the major subsystems for incineration and gasification technologies appear to be similar, the unit operations and fundamental chemical reactions that occur within each major subsystem are very different, perhaps with the exception of waste preparation. Each of these major process subsystems are described in more detail in the following paragraphs. Major emission and byproduct streams are identified, and unit operations within each major subsystem compared and contrasted.

2.1   Waste Preparation and Feeding

2.1.1 Incineration

The type of waste feed system for incinerators depends on the physical form of waste. Liquid wastes are blended and then pumped into the combustion chamber through nozzles to atomize the liquid feed. Liquid feeds may be screened to remove suspended particles that can plug the atomization nozzles. Blending is also used to control waste properties such as heating value and chorine content. Sludges are typically mixed and fed using cavity pumps and water-cooled lances. Bulk solids are shredded to obtain a more uniform particle size in the combustion chamber. Shredded solids are typically fed using rams, gravity feed, air lock feeders, screw feeders, or belt feeders (1).

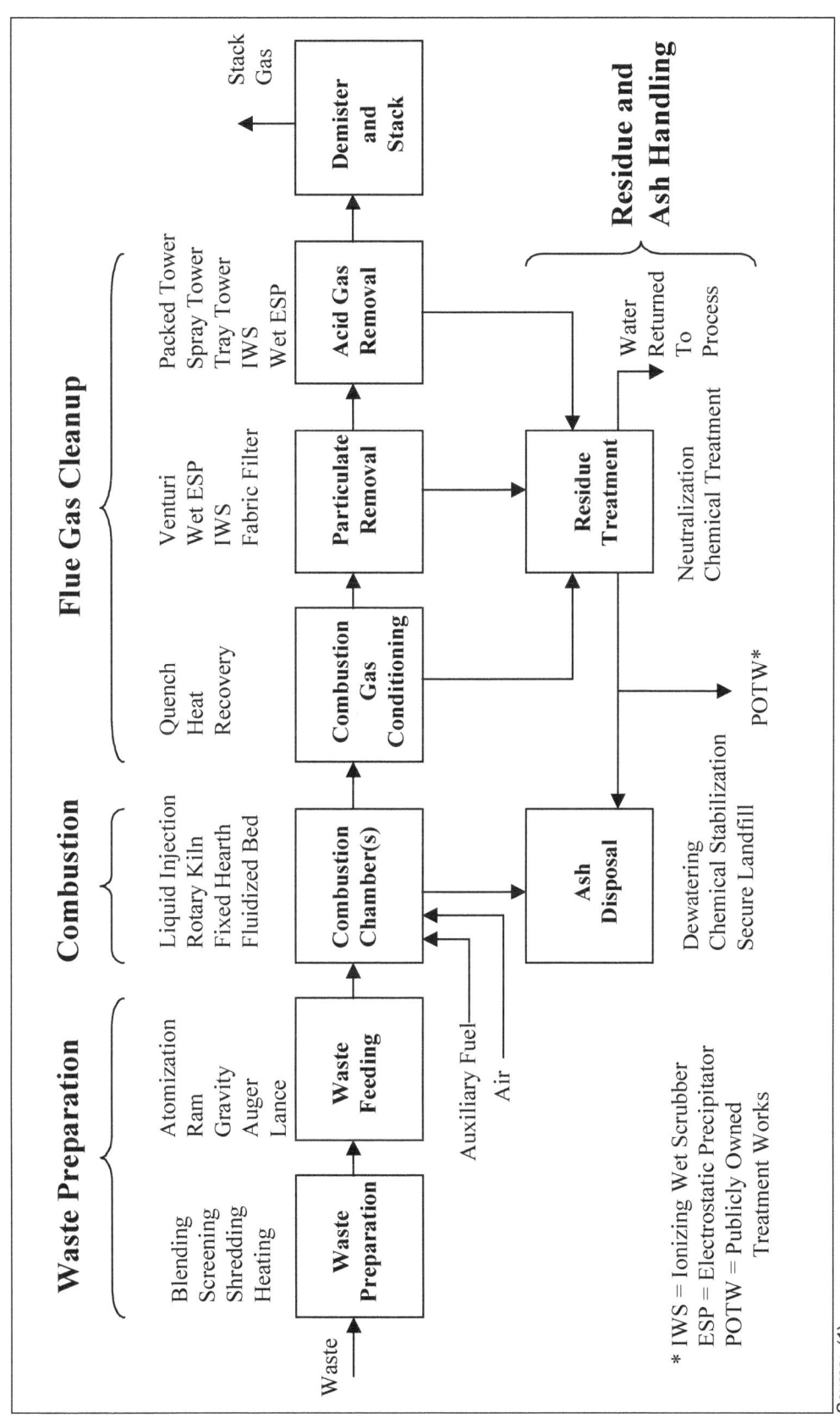

Figure 2-1. Incineration Process Flow Diagram

* IWS = Ionizing Wet Scrubber
ESP = Electrostatic Precipitator
POTW = Publicly Owned
        Treatment Works

Source: (1)

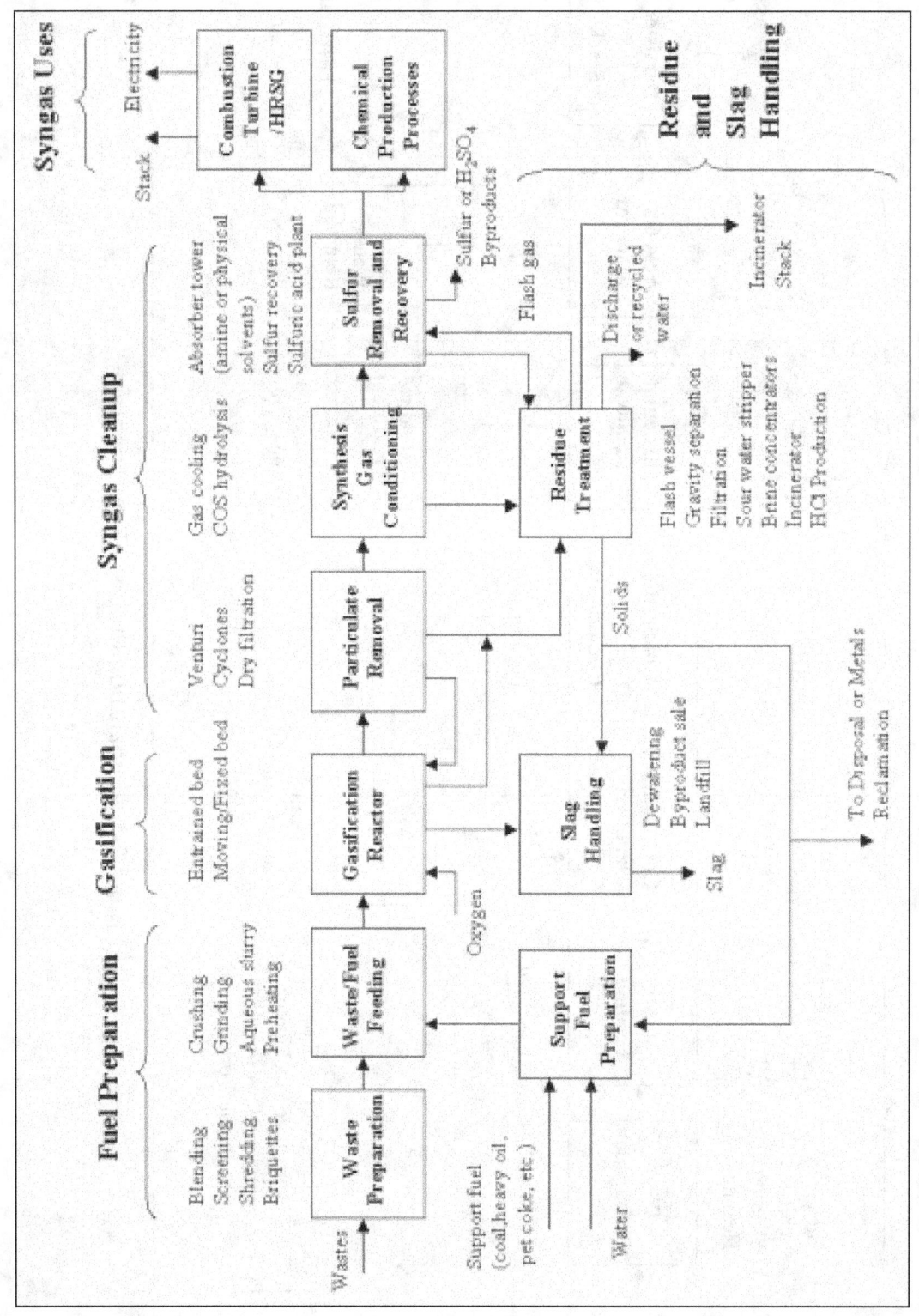

Figure 2-2. Gasification Process Flow Diagram

2.1.2 Gasification

In the gasification processes, fuel can be fed to the gasifier in the form of an aqueous slurry, dry solids, or liquids. Slurry and liquids are fed using high-pressure, positive displacement charge pumps in an enclosed system. Dry solids are pneumatically conveyed with nitrogen and fed through enclosed lockhoppers in the form of ground solids, pellets, or briquettes. Solid support fuels such as coal or petroleum coke are crushed and ground to the appropriate size before being gasified. For slurry fed processes, the ground solids are mixed with water (typically recycled from the process) in a wet rod mill to form an aqueous slurry. Primary fuel handling systems such as storage piles, conveyors, crushing, grinding, etc. are similar to systems used in conventional power systems and include unit operations for control of fugitive dust emissions.

Processes used for waste handling and preparation are similar to those used in the incineration industry or in the handling of secondary materials used for feedstocks in refinery cokers. Specific techniques depend on the physical form of the waste. Wastes can be combined with the support fuel before, during or after the fuel preparation process. For example, waste gasification tests were conducted in 1994 as part of EPA's SITE program (2). In this test program, a mixture of contaminated soil from the Purity Oil Sales superfund site, clean soil spiked with SAE 30 motor oil, and Pittsburgh #8 coal were gasified to demonstrate the process for destruction of a RCRA hazardous waste. Contaminated soil was transferred from drums into a waste feed hopper and metered into the wet rod mill along with the crushed coal using a bin feeder and bucket system to form an aqueous slurry. The solids grinding and slurry preparation unit included a baghouse and dust control system to control particulate emissions. Enclosed conveyor belts and coal handling equipment operated under slightly negative pressure. Particulate matter was collected in the baghouse and recycled to the fuel preparation process. The wet rod mill and slurry storage tank were enclosed and the vent gases, along with gases from the baghouse, were routed to a carbon canister for removal of organic compound vapors.

At the El Dorado refinery in Kansas, refinery RCRA hazardous wastes such as API separator bottoms (K051), acid soluble oils (D001, D018), primary wastewater treatment sludge (F037 and F038), and phenolic residue can be gasified in a dilute (2-5%) blend with petroleum coke (3, DelGrego Conference paper). At this facility, the coke slurry is prepared in a wet rod mill and the oily refinery wastes are blended in a second liquid feed system. The slurry and oily liquid feeds are fed to the gasifier using a single gasifier feed injector. The liquid feed system is designed so that it can be turned on and off while the gasifier is operating.

2.2 Combustion vs. Gasification

2.2.1 Incineration

Four major types of combustion chamber designs are used in modern incineration systems: liquid injection, rotary kiln, fixed hearth, and fluidized bed. Boilers and industrial furnaces (BIF units) are also examples of incineration systems; however, according to EPA MACT information less than 15% of the hazardous waste is disposed of in these units. The application of each type of combustion chamber is a function of the physical form and ash content of the wastes being combusted. In each of these designs, waste material is combusted in the presence of a relatively large excess of oxygen (air) to maximize the conversion of the hydrocarbon-based wastes to carbon dioxide and water. In some configurations, excess fuel and oxygen must be added to increase incineration temperatures to improve destruction and removal efficiency. This also increases the production and emission of carbon dioxide.

Sulfur and nitrogen in the feedstock are oxidized to form SO_x and NO_x. Halogens in the feedstock are primarily converted to acid gases such as HCl and HF and exit the combustion chamber with the combustion gases. Temperatures in the refractory-lined combustion chambers may range from 1200°F to 2500°F with mean gas residence times of 0.3 to 5.0 seconds (1,4).

Incinerators typically operate at atmospheric pressure and temperatures at which the mineral matter or ash in the waste is not completely fused (as slag) during the incineration processes. Ash solids will either exit the bottom/discharge end of the combustion chambers as bottom ash, or as particulate matter entrained in the combustion flue gas stream.

Liquid injection combustion chambers are used primarily for pumpable liquid wastes that are injected into burners in the form of an atomized spray using spray nozzles. Axial, radial, or tangential burner and nozzle arrangements can be used. Good atomization of the liquid waste feed is essential to obtain high destruction efficiencies in the combustion chamber.

Rotary kiln incinerators are used for a wide variety of feedstocks, including solids wastes, slurries, liquids, and containerized wastes. Combustion typically occurs in two stages; the rotary kiln and the afterburner. The rotary kiln is a cylinder which in mounted at a slight incline. As the cylinder rotates, waste material is mixed and transported through the combustion chamber where wastes are converted to gases through a series of volatilization, destructive distillation, and partial combustion reactions. The gas phase combustion reactions are then completed in the afterburner where operating temperatures may range from 2000°F to 2500°F. Liquid wastes are sometimes injected into the afterburner section to obtain additional waste destruction.

2-6

Fixed hearth incinerators also use a two-stage combustion process, much like rotary kiln systems. Unlike rotary kiln system, however, the waste is combusted under starved air conditions in primary stage where the volatile fraction is destroyed pyrolytically. Pyrolysis is the condition in which there is insufficient oxygen to react with all of the carbon in the feedstock, resulting in unburned carbon residual (soot). Temperatures in the first stage range from 1200°F to 1800°F. The starved air conditions minimize the amount of particulate entertainment and carryover into the combustion gases. The smoke and pyrolytic products then enter the secondary stage where the combustion process is completed using a large quantity of excess air.

Fluidized bed incinerators can be either circulating or bubbling bed designs. They are used primarily for incineration of sludge or shredded materials. In both systems, the combustion vessel contains a bed of inert particles (sand, silica, etc.) which is fluidized (bubbling bed) or entrained (circulating bed) using combustion air which enters the bottom of the vessel. In entrained bed systems, air velocities are higher such that solids are carried overhead with the combustion gases, captured in a cyclone and recycled to the combustion chamber. Operating temperatures are typically 1400°F to 1600°F. These systems also offer the option for in-situ acid gas neutralization within the fluidized bed by adding lime or limestone solids.

2.2.2 Gasification

Gasification is a thermal chemical conversion process designed to maximize the conversion of the carbonaceous fuel and waste to a synthesis gas (syngas) containing primarily carbon monoxide and hydrogen (over 85%) with lesser amounts of carbon dioxide, water, methane, argon, and nitrogen. The chemical reactions take place in the presence of steam in an oxygen-lean reducing atmosphere, in contrast to combustion where reactions take place in an oxygen-rich, excess air environment. In other words, the ratio of oxygen molecules to carbon molecules is less than one in the gasification reactor. The following simplified chemical conversion formulas describe the basic gasification process:

$$C(fuel) + O_2 \rightarrow CO_2 + heat \qquad \text{Reaction 2-1 (exothermic)}$$

$$C + H_2O(steam) \rightarrow CO + H_2 \qquad \text{Reaction 2-2 (endothermic)}$$

$$C + CO_2 \rightarrow 2CO \qquad \text{Reaction 2-3 (endothermic)}$$

$$C + 2H_2 \rightarrow CH_4 \qquad \text{Reaction 2-4 (exothermic)}$$

$$CO + H_2O \rightarrow CO_2 + H_2 \qquad \text{Reaction 2-5 (exothermic)}$$

$$CO + 3H_2 \rightarrow CH_4 + H_2O \qquad\qquad \text{Reaction 2-6 (exothermic)}$$

A portion of the fuel undergoes partial oxidation by precisely controlling the amount of oxygen fed to the gasifier (Reaction 2-1). The heat released in the first reaction shown above provides the necessary energy for the primary gasification reaction (Reaction 2-2) to proceed very rapidly. Gasification temperatures and pressures within the refractory-lined reactor typically range from 2200°F to 3600°F and near atmospheric to 1200 psig, respectively. At higher temperatures the endothermic reactions are favored. A wide variety of carbonaceous feedstocks can be used in the gasification process including: coal, heavy oil, petroleum coke, orimulsion, and waste materials (e.g., refinery wastes, contaminated soils, chlorinated wastes, municipal sewage sludge, etc.). Low-Btu wastes may be blended with high-Btu content supplementary fuels such as coal or petroleum coke to maintain the desired gasification temperatures in the reactor. However, unlike incineration, these supplementary fuels contribute primarily to the production of more syngas and not to the production of CO_2.

The reducing atmosphere within the gasification reactor prevents the formation of oxidized species such as SO_2 and NO_x. Instead, sulfur and nitrogen (organic-derived) in the feedstocks are primarily converted to H_2S (with lesser amounts of COS), ammonia, and nitrogen (N_2). Trace amounts of hydrogen cyanide may also be present. Halogens in the feedstock are converted to inorganic acid halides (e.g., HCl, HF, etc.) in the gasification process. Acid halides are easily removed from the syngas in downstream syngas cleanup operations.

The concentrations of H_2S, COS, HCl, N_2, and NH_3 in the raw syngas are almost entirely dependent on the levels of sulfur, chlorine, and nitrogen present in the feedstock, whereas the proportions of CO, H_2, CO_2, and CH_4 are indicators of gasifier temperature and oxygen:carbon:hydrogen ratios. In fact the methane concentration in the syngas has often been used as an operating control parameter with real-time process feedback available from on-line gas chromatographs or mass spectrometers.

Modern gasification systems, that meet the GTC definition of gasification as presented above, are applicable to refinery operations. These gasification systems can be categorized as entrained bed and moving bed (also known as fixed bed). Oxygen blown, high-temperature entrained gasification systems do not produce any tars or heavy oils. Fixed bed gasifiers can produce heavy oils and tars which are typically separated from the syngas and recycled to the gasifier. The higher temperatures promote higher carbon conversion rates than those found in many low-temperature, air-blown systems. Trace elements and metals in the feedstock are typically concentrated and immobilized in the glassy slag. A portion of the more volatile metals remains in the raw syngas and is captured in the downstream gas cleanup systems.

Entrained Bed

Several entrained-bed reactors equipped with either water quench or waste heat recovery systems are currently in use. In entrained bed gasifiers, fuel and oxygen enter the reactor in concurrent flow arrangements and in an appropriate ratio such that the gasifier is operating in a slagging mode (i.e., the operating temperature is above the melting point of the ash). In two-stage entrained gasifiers, additional fuel (in slurry form) is added to a second gasification stage to cool and enhance the heating value of the syngas from the first gasification stage. The molten ash flows into a water bath or spray at the exit of the gasifier. This process serves to solidify the molten ash, creating a glassy vitrified solid slag or frit material that is removed from the gasifier, either intermittently via a lockhopper system or through a continuous pressure letdown system. In quench gasifiers, the syngas is extracted with the slag and is cooled when it contacts the pool of water within the slag quench zone of the gasifier. Gasification units produce only a small amount of slag if the feedstock contains small amounts of heavy mineral matter.

Water from the quench chamber contains fine particulate, dissolved sulfur species, ammonia, and other water-soluble gases and is processed in a series of treatment steps as discussed later in this section. Other gasification systems without direct quench use waste heat recovery systems to cool the syngas downstream of the gasifier and produce steam that can be used for other process needs or for energy production in a steam turbine. A similar inert glassy slag is produced in this type of system.

Moving Bed (Fixed Bed)

In the moving bed gasifier, sized fuel (e.g., briquettes or pellets) is fed to the top of the gasifier. At the bottom, oxygen and steam enter and the slag is withdrawn. Liquid wastes can also be introduced into the gasifier at the bottom of the reactor vessel. As the solid fuel moves down through the bed, counter-currently to the rising syngas, it proceeds through four zones: drying, devolatilization, gasification and combustion. Drying occurs when the hot syngas contacts the feed at the top of the gasifier. Next the fuel devolatilizes, forming tars and oils. These compounds exit with the raw syngas, and are captured in downstream cleanup processes and recycled to the gasifier. The devolatilized fuel then enters the higher temperature gasification zone where it reacts with steam and carbon dioxide. Near the bottom of the gasifier the resulting char and ash react with oxygen creating temperatures high enough to melt the ash and form slag. The slag is then removed and quenched with water.

2.3   Flue Gas Cleanup vs. Syngas Cleanup

2.3.1 Incineration

Combustion gases from hazardous waste incineration systems are typically processed in a series of treatment operations to remove entrained particulate matter and acid gases such as HCl and other inorganic acid halides. Systems that process low ash, low halogen content liquid wastes may not require any downstream process controls. However, one of the more common gas cleanup configurations used at waste incineration facilities is a gas quench (gas cooling), followed by a venturi scrubber (particulate removal) and a packed tower absorber (acid gas removal). Wet electrostatic precipitators and ionizing wet scrubbers are used at some facilities for combined particulate and acid gas removal. Fabric filter systems are also used for particulate removal in some applications. Demisters are often used to treat the combustion gases before they are discharged to the atmosphere to reduce the visible vapor plume at the stack.

2.3.2 Gasification

Syngas from the gasification process is also treated in a series of gas cleanup and byproduct recovery operations. However, unlike incineration where combustion gases are treated at atmospheric pressure, the volume of syngas that must be treated in a gasification process is reduced significantly because of the elevated pressure of the syngas. Some of the operations such as gas quenching and/or heat recovery and particulate removal are similar to those used in incineration systems. Like incineration systems, wet scrubbers and dry filtration systems are often used to remove particulate matter and acid gases from the raw syngas. With highly chlorinated feedstocks, the hydrogen chloride can be recovered and used or sold as hydrochloric acid byproduct. However, this is where the similarities end. As discussed above, the chemical composition of the syngas is vastly different from that of combustion gases from incineration systems, and subsequent syngas treatment operations are designed to recover marketable byproducts.

After particulate matter is removed, the syngas is processed in a series of gas cooling steps where moisture, ammonia, and other water-soluble gas species are removed. The conditioned syngas then enters the sulfur removal and recovery process designed to remove H_2S and sometimes COS. These reduced sulfur species are recovered as elemental sulfur, or in some cases, converted to a sulfuric acid byproduct. The typical sulfur removal and recovery processes used to treat the raw syngas are the same as commercially available methods used in other industrial applications such as oil refining and natural gas recovery. One commonly used process to remove sulfur compounds is the selective-amine technology where reduced sulfur species are removed from the syngas using an amine-based solvent in an absorber tower. The rich solvent is regenerated in a stripper tower and circulated to the absorber. Physical solvents such as Selexol™, Rectisol™, and Purisol™ are also used. The reduced sulfur species removed in the solvent stripper are then converted to elemental sulfur in a sulfur recovery process such as the Selectox™/Claus process. Sulfur recoveries from H_2S are typically 95 to 99% (2, 3). The tail gas from the sulfur recovery unit contains low levels of sulfur compounds and can be treated in a cleanup unit (e.g., incinerator) or recycled back to the gasification unit to obtain overall sulfur recovery levels greater than 99%.

2.4   Residue and Ash/Slag Handling

2.4.1 Incineration

Ash is typically quenched with water or air cooled after discharge from the combustion chamber. The ash is accumulated in drums or storage ponds prior to disposal in a permitted hazardous waste landfill. The ash may be dewatered or subject to chemical fixation prior to disposal. Residues are also generated from the combustion gas cleanup systems during gas quenching, particulate removal and acid gas absorption. These cleanup processes typically generate solid ash streams or aqueous streams containing fine particulate matter and absorbed acid gases. Trace levels of organic contaminants may also be present. Solid residues are handled with the ash from the combustion chamber. Aqueous streams are typically neutralized and discharged to settling ponds or processed in a chemical precipitation or other common wastewater treatment operation. Concentrated contaminants (settled solids, treatment sludge, etc.) from these processes are ultimately disposed of in a landfill. The treated water may be recycled to the gas cleanup processes or discharged to a POTW.

2.4.2 Gasification

Glassy vitrified slag in the slag quench zone of the gasifier is discharged at the bottom of the gasifier vessel into a collection system where the solids are dewatered and the water is recycled to the process. In some cases, the slag is further separated into a coarse and fine fraction to obtain certain byproduct specifications. The separated non-toxic slag can be stored on-site and subsequently sold or loaded directly into railcar or truck to the byproduct (coarse slag fraction) market and/or disposed in a nonhazardous landfill.

Residual streams from unit operations in the gasification and synthesis gas cleanup systems differ depending on the configuration of the facility. In general, however, residual process water streams, containing dissolved gases, dissolved minerals, and fine particulate matter, will be generated in the syngas quench system and particulate scrubber. Dry filtration systems offer the advantage of reduced complexity in the process water handling and treatment systems. Process water streams are typically processed in a flash vessel under slight vacuum to remove the dissolved gases. Fine solids are then settled of filtered from the water and the clarified water is recycled to the process. The collected solids may be disposed with the slag, recycled to the fuel preparation system to recover energy value in the gasifier, processed further for reclamation of metals, or disposed as a separate material, depending on the characteristics of the fine solids.

For gasification of heavy refinery residuals and petroleum coke, specialized metals recovery systems are often used to recover metals such as nickel and vanadium that are present in the feedstocks at high concentrations (6,7,8,9,10). In general, these systems are designed to concentrate and collect heavy metals in the particulate matter (i.e., ash and unburned carbon) removed from the raw syngas. The processes typically involve filtration of the particulate matter from the process water stream to obtain a filter cake enriched in metals such a nickel and vanadium. The filtrate is recycled to the gasification process. The filter cake, containing unreacted carbon, can then be "roasted" (i.e., oxidized) in a furnace to recover the energy content and to produce a valuable ash product enriched in elemental oxides such as vanadium pentoxide which can be sold for use in the metallurgical industry.

Aqueous condensate streams from the gas cooling section and minor additional aqueous streams from the sulfur removal and recovery systems are typically processed in a sour water stripper where the water is steam-stripped for removal of dissolved gases (primarily hydrogen sulfide, ammonia and carbon dioxide). Conventional waste water treatment systems or brine concentrators are also used at some facilities for additional treatment of aqueous residual streams. The treated water is then discharged and/or recycled to the process.

The sour water stripper overhead vapor and gases from a flash can be recycled back to the sulfur recovery unit or routed to an incinerator for destruction. Tail gas is also produced in the sulfur recovery process and can either be recycled to the sulfur recovery unit, recycled to the gasifier, or routed to a small tail gas incinerator for destruction of contaminants such as H_2S, COS, CO and NH_3. Incinerator stack gases are vented to the atmosphere.

2.5   Syngas End Uses

The clean product syngas exiting the sulfur removal process has many potential uses. The syngas may be combusted in a gas turbine or gas turbine/combined-cycle (gas turbine with a heat recovery steam generator) power block to produce electricity and steam. Carbon monoxide and hydrogen are basic chemical building blocks for production of many chemicals. Thus, syngas may also be used as a feedstock in downstream chemical production processes at chemical plants or refineries. When hydrogen is a desired product, which is the case in many refinery gasification applications, the syngas can be reacted with steam to convert the carbon monoxide to hydrogen via the steam shift reaction:

$$CO + H_2O \rightarrow CO_2 + H_2$$

Other products that can be manufactured from syngas include: methanol, synthetic natural gas (SNG), fertilizers, isobutylene, methyl tertiary butyl ether (MTBE), acetic anhydride, tertiary amyl methyl ether (TAME), oxo alcohols, carbon dioxide, ammonia, formaldehyde, acetaldehyde, acetic acid and isobutanol (5). Methanol is the basic parent chemical for many of these compounds.

2.6   References

1. Dempsey, C., and Oppelt, T.E. "Incineration of Hazardous Waste: A Critical Review Update," *Air and Waste*, Vol. 43, January 1993.

2. U.S. EPA, "Texaco Gasification Process Innovative Technology Evaluation Report," Office of Research and Development Superfund Innovative Technology Evaluation Program, EPA/540/R-94/514, July 1995.

3. Rhodes, A.K, "Kansas Refinery Starts Up Coke Gasification Unit," *Oil & Gas Journal*, August, 5 1996.

4. DelGrego, G., "Experience with Low Value Feed Gasification at the El Dorado, Kansas Refinery," Presented at the 1999 Gasification Technologies Conference, San Francisco, CA, October 17-20, 1999.

5. U.S. EPA MACT Rule for Hazardous Waste Incinerators, 61 FR 17357, April 19, 1996.

6. Liebner, W., "MGP-Lurgi/SVZ Mulit Purpose Gasification, Another Commercially Proven Gasification Technology," Presented at the 1999 Gasification Technologies Conference, San Francisco, CA, October 17-20, 1999.

7. De Graaf, J.D., E.W. Koopmann, and P.L. Zuideveld, "Shell Pernis Netherlands Refinery Residue Gasification Project," Presented at the 1999 Gasification Technologies Conference, San Francisco, CA, October 17-20, 1999.

8. The Shell Gasification Process. Vendor literature process description. Shell Global Solutions U.S. Houston, TX, October 1999.

9. Maule, K. and S. Kohnke, "The Solution to the Soot Problem in an HVG Gasification Plant," Presented at the 1999 Gasification Technologies Conference, San Francisco, CA, October 17-20, 1999.

10. Heaven, D.L, "Gasification Converts a Variety of Problem Feedstocks and Wastes," *Oil & Gas Journal*, May 1996.

3.0   Byproduct Treatment and Utilization

This section provides more detailed information regarding specific byproduct and emission streams from gasification and incineration processes and their possible utilization or treatment. An overview of the auxiliary systems designed to recover or treat the byproducts from both technologies is provided in Table 3-1. The numerous auxiliary treatment systems and broader applications of byproducts suggest the gasification process is an intermediate stage in a refining process where the gasifier breaks down low-value complex materials into simple and useful components. Multiple auxiliary systems separate and recover these byproducts for the production of more valuable commodities. Both gasification and incineration technologies are capable of producing the commodities of heat, steam, and electric power, and the potential for metals or acid reclamation, however, the incineration process is the final treatment process leading to the direct production of emissions streams which require additional treatment prior to disposal.

3.1   Byproducts of Gasification

3.1.1 Slag/Vitreous Frit

Slag or vitreous frit is the primary solid byproduct of gasification. The slag contains the mineral matter associated with the feed materials in a vitrified form, a hard, glassy-like substance. This is the result of gasifier operation at temperatures above the fusion, or melting temperature of the mineral matter. Under these conditions, non-volatile metals are bound together in a molten form until it is cooled in a pool of water at the bottom of a quench gasifier, or by natural heat loss at the bottom of an entrained bed gasifier. Volatile metals such as mercury, if present in the feedstock, are typically not recovered in the slag, but are removed from the raw syngas during cleanup.

Slag production is a function of how much mineral matter is present in the gasifier feed, so materials such as coal produce much more slag than petroleum feedstocks. Regardless of the feed, as long as the operating temperature is above the fusion temperature of the ash (true for the modern gasification technologies under discussion), slag will be produced. Its physical structure is sensitive to changes in operating temperature and pressure and, in some cases, physical examination of the slag's appearance can provide a good indication of carbon conversion in the gasifier.

Table 3-1. Byproduct Treatment and Utilization: Gasification vs. Incineration

Subsystem	Function	Byproducts	Characteristics	Additional Treatment	Potential Uses	Disposal
			Gasification Process			
Gasifier Lockhopper and Quench	Slag removal and gas cooling (quench)	Slag	Glassy, vitrified solid - Usually passes all TCLP parameters	Dewatering; May be sorted by particle size	Metals reclamation; road bed filler; abrasives	Landfill
		Grey Water	Quench system water or water from gasification	Clarification and recycle to gasifier		WWT
		Raw Syngas	Medium-Btu gas composed of CO, H_2, CO_2, H_2O, N_2, Ar, CH_4, HCl and H_2S. Minor components include COS, NH_3, HCN, and other hydrocarbons.	Particulate removal, gas conditioning, and sulfur removal		Emergency Flare only
Particulate Removal System (dry filtration, clarifier, high-pressure solids settler, venturi scrubber, etc)	Removal of fine particulate matter	Ash/Char	Unconverted carbon and lighter slag particles	Separated from Black Water by filtration, naphtha extraction, or gravity. None required for dry removal systems.	Recycled to gasifier, metals reclamation	
		Black water (wet removal systems only)	Water containing particulate matter and water-soluble gaseous components	Filtration or decanter to remove solids; flash separator prior to WWT	Recycled to gasifier, processed for recovery of HCl	WWT
		Flash gas	Dissolved gases	Sulfur removal		Incineration
Gas Conditioning	Gas cooling and moisture removal	Sour water	Syngas condensate saturated with NH_3 and CO_2 and H_2S	Steam stripping to remove dissolved gases		Recycle to gasifier
		Sour gas	Primarily NH_3, CO_2, and H_2S	From steam stripper	NH_3 and CO_2 Recovery	Incineration or recycle

Table 3–1 (continued)

Subsystem	Function	Byproducts	Characteristics	Additional Treatment	Potential Uses	Disposal
Gasification Process (continued)						
Solvent scrubber	Sulfur removal	Acid gas	High H_2S and CO_2 concentration	Sulfur recovery system (Claus)	Elemental sulfur recovery, H_2SO_4 production	Tail gases from Claus system are incinerated
		Spent solvent	Amine-based or physical solvent	Steam stripped to remove acid gases and recycle the solvent	Solvent is regenerated and recycled	
		Heat-stable salts and other solids	Heat-stable salts are those that result from reaction of other acid gases with the amine that are not removed by steam stripping.	Solvent treated with NaOH or other system (e.g. electrolytic regeneration) to remove salts and filtration to remove non-soluble solids		
		Clean Syngas	Medium-Btu gas composed of CO, H_2, CO_2, N_2, Ar, and CH_4 with trace levels of other hydrocarbons and acid halides.	As required for specific application.	Gas turbine generator, hydrogen shift conversion, LP-MeOH process, gas separator column, etc.	Emergency Flare only
Incineration Process						
Ash Removal Systems	Removal of bottom ash	Bottom ash	Ash solids	Dewatering; Chemical fixation/stabilization		Hazardous waste land disposal
		Quench water		Settling, neutralization		POTW
Particulate Removal System (Wet ESP, Venturi Scrubber, Filter, IWS)	Removal of fine particulate matter	Ash solids	Fine particulate matter			Hazardous waste land disposal
		Scrubber water		Settling, neutralization	Recycle to process	POTW
Acid Gas Removal System (Absorber)	Removal of halide acid gases	Absorber water	Contains absorbed acid gases	Neutralization	Recovery of haloacids (HCl); Recycle to process	POTW

Because the slag is in a fused, vitrified state, it rarely fails the TCLP protocols for metals. Slag is not a good substrate for binding organic compounds so it is usually found to be nonhazardous, exhibiting none of the characteristics of a hazardous waste. Consequently, it may be disposed of in a nonhazardous landfill, or sold as an ore to recover the metals concentrated within its structure. Slag's hardness also makes it suitable as an abrasive or road-bed material as well as an aggregate in concrete formulations.

3.1.2 Fine Particulate Matter

Downstream of the quench system or syngas cooler, any unconverted carbon fines (char) and light ash material are scrubbed from the syngas with water in a venturi scrubber or removed using a dry filtration system. The solids are recovered from the scrubber water by a variety of techniques including filtration in a filter press, extraction with naphtha in a decanter, and by separation in a solids-settling vessel. Because the fines typically contain a high percentage of carbon, they are usually recycled to the gasifier, however sometimes the char/ash solids are collected and treated in a metals recovery process or sent to a metal-ore processing facility to reclaim the metals.

The decanter process is common to petroleum refinery gasifiers where gasifier temperatures may run too hot when carbon conversion is maximized. At higher operating temperatures, the excessive erosion of the refractory material lining the gasifier vessel may not be worth the incremental conversion rate. To compensate, a lower temperature and slightly lower conversion rate is acceptable based on the recovery and recycling of the resulting char in a decanter. In the decanter, the scrubber water and char are added to naphtha. The naphtha is not miscible in the water so it forms a separate layer on top of the water. The carbon (soot) in the char has a strong affinity for the naphtha layer so it migrates to the organic layer where it is decanted from the water. The naphtha-char stream is then added to fresh charge oil (typically vacuum distillation residuals or heavy oil) and sent to a stripping tower where the lighter naphtha is distilled from the oil before being recycled back to the decanter. The charge oil and carbon are then fed back to the gasifier. Ash material tends to stay with the water layer which is recycled back to the scrubber or blown down to a waste water treatment unit where the ash is recovered in the form of a filter cake. In refinery applications, the filter cake typically contains high concentrations of nickel and vanadium from the heavy residue feedstock and can be processed further for metals reclamation. In other processes, the fine char material in the scrubber water is filtered and the resulting filter cake combusted under controlled conditions to recover the energy value from the unreacted carbon and convert the metals such as vanadium to metal oxides such a vanadium pentoxide which is a valuable byproduct for metallurgical industries. The resulting ash may contain as much as 35 weight percent vanadium.

The fines entrained in the syngas, especially fines from entrained flow gasifiers, typically exhibit an enrichment in volatile metals similar to that associated with pulverized coal combustion in steam generating utility boilers. In addition to the mechanism of condensation on the large surface area of fine particles, the fine carbon acts as an adsorbent also provides a mechanism for enrichment. Unlike the slag material, these high-carbon containing solids can occasionally fail the TCLP characteristic for metals such as lead. The value of these solids through metals reclamation or recycling of carbon makes the disposal of this stream an unattractive option economically, but like any other solid waste product, it must be tested by the TCLP and for other hazard characteristics if disposed.

Fixed bed gasifiers do not experience the same degree of particulate or fines carryover as the entrained flow technologies. Further, fines from a fixed bed unit are readily returned to the gasifier. Residual solids collected in the condensate from gas cooling are ultimately returned to the bottom of the gasifier so that the material can be included in the vitreous slag product.

3.1.3 Process Water

Within synthesis gas conditioning and particulate removal steps, if wet particulate removal is utilized, different water streams result which require treatment. One advantage of a dry filtration system is that these process water streams are not generated which simplifies subsequent water treatment requirements. The various water streams used to cool and clean the syngas are typically recycled to the feed preparation area, to the scrubber after the entrained solids have been removed, to a zero discharge water system, or to a wastewater treatment system. However, recycling of water has its limitations as dissolved salts accumulate to levels incompatible with the process or its metallurgy. Process water is partially exchanged with fresh make-up water as process water is blown down to a wastewater treatment facility prior to discharge. Zero-discharge process water systems have no wastewater discharges by design, however these systems must address the removal of salts as a reclaimed product from brine evaporation.

Since these scrubber waters and the gas condensate are saturated with the water-soluble components present in syngas, these water streams are typically high in dissolved solids and gases with the following ionic species commonly found: sulfide, fluoride, chloride, formate, ammonium, cyanide, thiocyanate, and bicarbonate. One method of treatment for these water streams offers an additional opportunity to recover sulfur. Process water taken directly from the high temperature and pressure systems can be "flashed" in a vessel at low or negative pressure to release the dissolved gases. The flash gas is routed to the sulfur removal unit with the raw synthesis gas, and the water is either recycled to the system or it is blown down to a conventional wastewater treatment unit before discharge.

Gas condensate, also known as sour water, may also be steam-stripped to remove ammonia, carbon dioxide, and hydrogen sulfide usually dissolved in the condensate while under system pressure. The stripper overhead containing these gases can be routed to the sulfur recovery unit or they may be incinerated, subject to permit limitations for NO_x and SO_2 emissions. The sour water stripper recovers water suitable for recycling back to the process as make-up water to the various gas scrubbing and feed systems. A portion of the recovered water from the sour water stripper may be discharged to a conventional waste water treatment system (e.g., flocculation/sedimentation followed by biological treatment is standard for most refineries).

3.1.4 Sulfur Removal System

Amine-based solvents are routinely used in refineries for the removal of H_2S from process gases. There are trade names and acronyms covering a wide variety of these solvents including Selectamine™, Ucarsol™, and Sulphinol™. Physical solvents such as Selexol™, Rectisol™, and Purisol™ are also used. These solvents all absorb acid gases such as H_2S, and to a lesser extent, CO_2 and COS from the syngas in an absorber tower. These dissolved gases are readily steam-stripped from the rich solvent in a stripper tower where they form a concentrated acid-gas stream that contains percent-levels of H_2S. The lean solvent is recycled back to the absorber in a closed loop.

The acid gas is treated in a Claus unit or other sulfur production unit under controlled conditions to produce elemental sulfur that is condensed and stored in molten form in a steam-traced or heated vessel, or cooled and stored in solid cake/powder form prior to sale and transport. This is a high quality sulfur with excellent market value. Tail gas from the sulfur unit is either incinerated, returned to the acid gas removal step for reprocessing, or recycled to the gasifier subject to operating permit restrictions.

The high levels of sulfide typically present within this system affect the chemical and phase equilibria of various substances in the syngas. Any vapor-phase metals present in the syngas that form insoluble metals sulfides are likely to be precipitated as solids in the recirculating solvent. Volatile metals like mercury that are not captured in the slag may be collected by the metal-sulfide mechanism in the sulfur removal system. If the metal sulfides are not volatile enough to be stripped out with the H_2S, they are likely to accumulate in the solvent where they are removed by filtration or solvent exchange.

In addition to metals, other acidic components not effectively removed by the scrubber water systems may bind to physical separation solvent or to the amine solvent in an acid-base neutralization reaction that creates a heat-stable amine salt. These compounds are stable because the salt does not decompose during the steam stripping process. This accumulation reduces the effectiveness of the solvent so the solvent must be regenerated periodically with a strong alkali to free the amine. The regeneration is usually done by the solvent vendor on a periodic basis, however sometimes a smaller continuous unit is included as part of the normal plant operation.

3.1.5 Clean Syngas Product

The major gaseous stream produced in the gasification process is the clean syngas product. If the syngas is combusted in a gas combustion turbine for power production, the combustion gases exiting the turbine and heat recovery steam generator (HRSG) become the major gaseous emissions. Gaseous emissions for downstream use of the product syngas in chemical production processes will vary depending on the process and products produced. When utilizing the syngas for chemical production very pure syngas is required before chemical processing so the syngas is not expected to contribute to downstream emissions. A detailed discussion of gaseous emissions from these chemical production processes is considered outside the scope of this paper; however, all such facilities would be covered by current regulatory oversight. Therefore, subsequent discussions in this paper will focus on the use of syngas for energy production.

3.2 Byproducts of Incineration

3.2.1 Ash

Bottom ash and fly ash are the two primary solid byproducts from incineration. Bottom ash exits the combustion chamber and is either air-cooled or quenched with water. The ash is usually accumulated on site prior to disposal in a hazardous waste landfill. In some cases, it may be dewatered or chemically stabilized to meet land disposal restrictions.

The entrained fly ash is removed from the flue gas by air pollution control devices. The most common system for removal of fly ash, acid gases, and other contaminants is a quench system for gas cooling, followed by a venturi scrubber for particulate removal, and a packed absorber for acid gas removal. The water streams from these systems containing the ash solids, absorbed acid gases, salts, and traces of organic compounds are collected in sumps or settling tanks. At this point, solids are settled and acids are either neutralized or collected before the water is recycled to the system or discharged to a POTW.

Because the ash is produced in an oxidizing environment, the ash solids are composed primarily of elemental oxides. Volatile and semi-volatile metals are typically found enriched in the fine particles by the same mechanism described earlier or elsewhere for coal combustion systems. The RCRA rules defining ashes from listed hazardous wastes as having the same hazardous waste prohibits general landfill disposal. This "derived from" rule is being challenged in favor of other restrictions based on the TCLP hazard characteristic tests.

3.2.2 Process Water

Water from the quench system, and other air pollution control devices is recycled to the process whenever possible following solids removal and neutralization. However, a portion of the water must eventually be blown down to avoid accumulation of salts and other contaminants. Various treatment options may be used to recover salts and acids, but ultimately water is discharged to a wastewater treatment facility.

4.0 Regulatory and Environmental Concerns

4.1 Regulatory Issues

The Resource Conservation and Recovery Act (RCRA) governs the handling of hazardous materials; classified as either "listed" or "characteristic" waste. Listed wastes are those materials identified in 40 CFR Part 261 Subpart D. Characteristic wastes include other, non-listed substances that meet the characteristics of a hazardous waste defined in 40 CFR Part 261 Subpart C. According to RCRA, materials produced from, or in contact with a hazardous waste, themselves become hazardous wastes. For incinerators and presumably gasifiers (if regulated), the solid byproducts, wastewater streams, and gaseous products would be included in that definition and be regulated under RCRA.

However, the issue of when a hazardous material is considered a "waste" has often been challenged. The "function" a hazardous material serves within a process boundary has been used as an argument against the classification of that material as a waste. The EPA has not generally accepted the identification of a material as a fuel or intermediate stream for further refinement or recovery as an exemption from RCRA. As an example, the BIF rule was promulgated to cover the incineration of hazardous materials that were being used as fuels in heat recovery boilers and furnaces.

Nevertheless, certain exceptions and exclusions from RCRA have been granted. The most relevant are the petroleum coker (63 FR 42109) and comparable fuels exclusions (40 CFR 261.38). The syngas provision of the comparable fuels exclusion states that if the syngas produced from a hazardous waste meets the criteria in 40 CFR 261.38, then the exclusion would apply and the syngas would not be regulated as a solid waste. It is noteworthy that RCRA's jurisdiction over syngas fuels produced from hazardous waste is currently being challenged by gasification industry representatives. Unlike gasification, incineration does not produce a fuel gas and so this exclusion is not applicable to incineration.

In addition, RCRA contains specific performance standards for operation of hazardous waste incinerators. Facilities must demonstrate compliance with the following criteria by conducting a "trial burn" for the specific waste(s) that are to be incinerated:

- 99.99% DRE for each POHC in the waste feed (99.9999% for dioxin/furan or polychlorinated biphenyl listed wastes);

- At least 99% removal of HCl if HCl stack emissions are greater than 1.8 kg/hr; and

- Particulate matter emissions no greater than 180 mg/dscm at 7% oxygen.

The concept and selection of a POHC is an important part of the incineration regulations. POHCs for the trial burn assessment must be selected from the RCRA Appendix VIII list of over 450 substances, based on the POHCs that are present in the waste feed and are most difficult to incinerate. EPA frequently requires that site-specific risk assessments, incorporating direct and indirect exposures, be conducted during the combustion unit's permitting process.

EPA has also recently finalized MACT standards for the incineration of hazardous wastes as mandated by the 1990 CAAA (64 FR 52828). MACT standards have been established for both new and existing hazardous waste incinerators and include stack gas concentration limits for particulate matter, low volatile metals (Sb, As, Be and Cr), semi-volatile metals (Pb and Cd), mercury, dioxin/furan compounds, carbon monoxide, total chlorides (HCl/Cl_2), total hydrocarbons, and DREs. Specific limits are provided in Table 4-1.

Table 4-1. Final MACT Standards for Hazardous Waste Incinerators

HAP or HAP Surrogate	Existing Incinerators (7% O_2)	New Incinerators (7% O_2)
Dioxin/furans	0.20 ng/dscm TEQ, or 0.40 ng/dscm and temperature at inlet to the particulate control device $\leq 400°$ F	0.20 ng/dscm TEQ
Particulate Matter	34 mg/dscm (0.015 gr/dscf)	34 mg/dscm (0.015 gr/dscf)
Mercury	130 µg/dscm	45 µg/dscm
SVM	240 µg/dscm	24 µg/dscm
LVM	97 µg/dscm	97 µg/dscm
HCl + Cl_2	77 µg/dscm	21 µg/dscm
Hydrocarbons [a]	10 ppmv (or 100 ppmv CO)	10 ppmv (or 100 ppmv CO)
DREs	99.99% (99.9999% for dioxin-listed wastes)	99.99% (99.9999% for dioxin-listed wastes)

Hourly rolling average expressed as propane.
SVM = Semi-volatile metals (Cd, Pb).
LVM = Low volatile metals (Sb, As, Be, Cr).
DRE = Destruction and removal efficiency for each specific POHC, except 99.9999% for specific dioxin-listed wastes.

These rules appear to have one thing in common and that is they generally regulate the final stage of recovery that is a direct producer of gaseous emissions. One of the virtues of gasification is that it is never a direct producer of gaseous emissions. The syngas is always used for its heating value or chemical composition. It is the gaseous emissions from other processes utilizing the syngas as a fuel that are themselves regulated under the CAA and their individual operating and discharge permits.

Another virtue is that many of the systems that process syngas for removal and recovery of marketable products like sulfur are well-known and established processes in the petroleum refining and chemical industry. Like those industries, the application of the RCRA TCLP requirements for solids and Clean Water Act provisions for effluent water streams are appropriate.

4.2 RCRA Exclusions Applicable to Gasification

4.2.1 Petroleum Coker Exclusion

It is important to understand EPA's rationale for granting the petroleum coker exclusion for refineries, so that the implications of a similar exclusion for gasification of hazardous oil-bearing residuals can be evaluated. Key points in EPA's decision to grant the petroleum coker exclusion were originally discussed in the November 1995 proposed rule (60 FR 57747). In this proposed rule, EPA noted that in earlier evaluations of exclusions for the refinery industry, the Agency decided not to grant a RCRA exclusion for the insertion of hazardous oil-bearing residuals into petroleum cokers "because of concerns about the fate of the hazardous constituents that may be contained in the recovered oil." In this rule dated July 28, 1994, EPA limited the Recovered Oil Rule to "recovered oil from petroleum refining, exploration and production that are inserted into the petroleum refining process prior to distillation and catalytic cracking." Thus, this final exclusion did not apply to recovered oil reinserted into the petroleum coker and it specifically excluded the RCRA listed hazardous refinery wastes (K048-K052, F037, and F038). However, after promulgation of the recovered oil rule, EPA received numerous comments and additional data from petroleum industry representatives on the composition of oil-bearing refinery residuals and the fate of toxic constituents contained in the secondary materials that are typically inserted into the petroleum coker. After review of these additional data, EPA decided to broaden the recovered oil exclusion to include "all secondary oil-bearing materials that are generated in the petroleum refining industry and are inserted into the petroleum refining process (including distillation, catalytic cracking, fractionation, or thermal cracking [i.e., coking])" (63 FR 42109). This definition includes all of the listed hazardous wastes from refinery operations (K048-K052, F037, and F038). EPA limited this exclusion to the production of coke which does not exhibit one or more of the characteristics of a hazardous waste. If the final coke product does not exhibit these characteristics, then both the coke product and the secondary materials used to produce the coke product are excluded from regulation under RCRA.

This detailed review of the coking process convinced EPA that the coker was in fact an integral part of the petroleum refining process and is similar to other refining processes such as distillation and catalytic cracking. EPA concluded that the coker contributes significant revenue to the refinery primarily through upgrading of lower value hydrocarbons into light ends that are used to produce more valuable product fuels. The primary purpose of the coker, as EPA explained in the proposed rule, is to thermally convert longer-chain hydrocarbons to recover the more valuable middle and light end hydrocarbons that are used to produce high-grade fuels. The typical coker yield is about 25 to 30% petroleum coke and 70% light hydrocarbons that are returned to the refining process to produce high-grade fuels. EPA also reviewed additional data on the composition of oil-bearing hazardous sludges relative to crude oil residuals that are typically fed to the coker and found that oil-bearing sludges generated during the refining process are substantially similar to normal coker feedstock material. Based on these data, EPA concluded that recycling of these hazardous oil-bearing materials, which comprised only about 1 to 3% of the total amount of refining residuals that are typically fed to the coker, can be accomplished without raising heavy metals concentrations to levels of concern in the final product coke.

Particular metals of concern for the listed refinery wastes (K048-K052) include chromium and lead, since these substances are introduced into the refinery process and do not originate primarily in the initial crude oil. In EPA's response to comments on the proposed rule for the refinery coker exclusion, the Agency notes that the level of metals in sludges generated by petroleum refineries and typically fed to the coker are, for the most part, comparable to the concentrations of metals in normal refinery feedstocks (1). EPA concedes that the levels of chromium and lead in the listed refinery wastes may be higher than normal petroleum feedstocks; however, EPA notes that the concentrations of these two metals in petroleum sludges are expected to decrease due to changes in the production process. New NESHAP and MACT standards promulgated under the Clean Air Act will result in chromium no longer being used in cooling towers, thus eliminating the principal source of chromium contaminants in the production process. Lead levels in petroleum sludges will continue to decline due to the phasing out of leaded gasoline as a product line. Data provided to EPA showed that the other hazardous metals found in the hazardous secondary materials can be traced back to the metals found in the original crude oil feedstock so they do not represent contaminants introduced through means other than the continued processing of the raw materials. Thus, "EPA's traditional concerns about unnecessary hazardous constituents being processed and ending up within the product were mitigated in this case because EPA viewed coking as the continual processing of a raw material that contains hazardous constituents, with concentrations of constituents found in the feedstock streams varying depending on the point in the process."

EPA's proposal to broaden the refinery coker exclusion to include gasification of the same secondary oil-bearing materials within a refinery is a logical extension of this rule when viewed in the context of the refinery coker exclusion. Like coking, gasification can be viewed as an integral part of the refinery process where the hydrocarbons in the secondary oil-bearing residuals are recovered by chemically converting them into a useful syngas product analogous to the production of coke (fuel) and recovery/recycle of hydrocarbons in the coker process. The product syngas can be used for energy production or further processed to produce other chemicals (e.g., hydrogen recycled to the refining process, and/or for production of methanol, acetic anhydride, etc.).

4.2.2 Comparable Fuels Exclusion

A second regulatory exclusion that is applicable to the gasification of hazardous waste materials is the RCRA Comparable Fuels Exclusion as described in the final rule of June 19, 1998 (63 FR 33781). In the final rule, "EPA has excluded from the regulatory definition of solid waste derived fuels that meet specification levels comparable to fossil fuels for concentrations of hazardous constituents and for physical parameters." The goal of this exclusion was to assure that an excluded waste derived fuel is similar in composition to commercially available fuels and therefore poses no greater risk than burning fossil fuel. A specific provision of this exemption applies to syngas derived from hazardous waste "from the thermal reactions of hazardous wastes by a process designed to generate both hydrogen (H_2) and carbon monoxide (CO) as a usable fuel." Inclusion of this provision is important because in doing so, EPA has established that they have jurisdiction under RCRA to regulate syngas produced from hazardous waste. Under this exemption, syngas produced from hazardous waste is excluded from RCRA requirements if it meets the following specifications:

- A minimum Btu value of 100 Btu/scf;

- A total halogen content of less than 1 ppmv;

- A total nitrogen (other than diatomic nitrogen, N_2) content less than 300 ppmv;

- A hydrogen sulfide content less than 200 ppmv; and

- Less than 1 ppmv of each hazardous constituent in the target list of Appendix VIII constituents.

Appendix VIII constituents include over 450 organic and inorganic substances of concern, including various metals and metal compounds (Sb, As, Ba, Be, Cd, Cr, Pb, Hg, Ni, Se,

Ag, and Tl), acid gases, polycyclic organic compounds, volatile organic compounds, and halogenated organic compounds.

The comparable syngas fuel exclusion applies only if the fuel is burned in the following units (that are also subject to Federal, State and local air emission requirements, including all applicable CAA MACT standards): 1) industrial furnaces; 2) industrial boilers; 3) utility boilers; and 4) hazardous waste incinerators. Residuals resulting from the treatment of a hazardous waste listed in Subpart D of RCRA to generate a syngas fuel remain a hazardous waste. The exclusion also contains specific requirements for testing and measurement of hazardous constituents to verify that the syngas meets the specification and qualifies for the exemption. Waste analyses plans must be developed and written which describe the procedures for sampling and analysis of the syngas. These plans must be submitted to and approved by the appropriate regulatory authority before performing sampling, analysis, or management of a syngas fuel as an excluded waste.

Application of this exemption to gasification of listed refinery wastes would require facilities to sample and analyze the clean product syngas stream. Established sampling and analytical methods exist for Btu content, total halogens, total nitrogen and hydrogen sulfide, so measurement of these parameters should not present any major problems. However, sampling and analytical methods for many of the Appendix VIII compounds in a reduced syngas matrix may not be fully developed or validated. Verifying compliance with the 1 ppmv specification for these compounds could be difficult in some cases. The status of sampling and analytical methods for reduced syngas streams is discussed in more detail in Section 5.

4.3   References

1. U. S. EPA. Notice of Data Availability (NODA) Response to Comment Document. Part II. Office of Solid Waste, June 1998.

5.0 Discussion

There are not many sources of data available regarding the environmental performance of gasification on RCRA wastes, and whether there are any sources comparing incineration and gasification of the same waste materials is unknown. However, what is evident from the existing data is a compelling case in favor of the gasification of oil-bearing refinery wastes and other hazardous materials.

One of the most applicable data sets can be found in a Technology Evaluation Report prepared in 1995 by Foster-Wheeler Enviresponse, Inc. (FWEI) under the EPA Superfund Innovative Technology Evaluation (SITE) Program. The report presents an evaluation of the Texaco Gasification Process on a coal-soil-water fuel with chlorobenzene added as a POHC to measure the process' destruction and removal efficiency (DRE). Lead and barium salts were also added to track the fate of these and other heavy metals.

The SITE Report's evaluation concluded that the DRE for chlorobenzene was greater than 99.99% while producing a clean syngas comparable to that produced from coal-water slurry alone. It was also noted that the coarse slag passed the TCLP; however, the fine slag and clarifier solids failed to meet the TCLP regulatory limit for lead. This was explained as a consequence of partitioning the more volatile metal species on the fine slag and carbon particles exiting the gasifier. This is also consistent with findings at the CWCGP where the more volatile elements were associated with the finer slag particles and high carbon-containing char removed in the wet scrubber. Data from this SITE Report are included in the section on available data.

In the introduction of the SITE Report, there is a reference to a California Department of Health Services report on the successful gasification of hazardous waste materials from an oil production field. Results of this and other tests on materials such as municipal sewage sludge, coal-liquefaction residues, and surrogate contaminated soil (clean soil and unused motor oil) have been used as the basis for permit applications for other commercial facilities throughout the United States.

One such facility is the El Dorado, Kansas refinery. The El Dorado refinery started a coke gasification unit in 1996 and it has been operating successfully on petroleum coke with supplemental feeds accounting for approximately 10 tons per day of listed refinery wastes. Among the waste streams that can be gasified are API separator bottoms (K051), acid-soluble oils (ASO) from the alkylation unit (D001 or D018), primary waste water treatment sludge (F037 and F038), and phenolic residue. The syngas produced is used to supplement natural gas fed to a combustion turbine and it accounts for about one-third of the turbine's fuel capacity.

The El Dorado gasifier is the first such unit to process listed hazardous wastes without a RCRA Part B permit. The Kansas Department of Health & Environment (KDHE) and EPA agreed in May 1995 that a Part B permit was not required on the basis that the gasifier was considered a processing unit. Other contributing factors in this decision included the net reduction in NO_x and SO_x emissions resulting from the gasification rather than off-site incineration of the pet coke, and the decrease in emissions from outside utilities no longer needing to produce the 40 MW being generated from the gasifier-fed turbine.

The El Dorado site has set a precedent for using refinery waste products for fuel while producing environmentally acceptable and marketable byproducts and lower emissions. The unit boasts a carbon conversion rate of about 99% with only 1% of the coke's mass collected as slag. Other solids recovered in the gas scrubbing system are collected in the form of a filter cake and are recycled to the gasifier to reclaim the energy value. The solids are being considered for metals reclamation or sale as a low-grade fuel.

The sulfur contained in the feedstock is converted to H_2S and COS in the syngas. These sulfur compounds are recovered in the refinery's amine-based acid-gas absorber. From there, the H_2S is sent to the sulfur recovery unit where greater than 99% of the H_2S is converted to elemental sulfur.

Water from the quench chamber and wet scrubber is treated by flashing to remove dissolved gases which are combined with the acid gas stream on its way to the sulfur recovery unit. After fine solids are allowed to settle, most of the clarified water is then recycled back to the process. Excess water is sent to the existing refinery water treatment system and requires no specialized treatment systems. The only gaseous emissions are those from the combustion turbine-heat recovery steam generator exhaust and these emissions were found to be orders of magnitude lower than those produced from the direct combustion of petroleum coke.

These beneficial factors would be consistent with gasification of any materials that would ordinarily be disposed of by incineration. Based on the performance at El Dorado, it appears reasonable to consider an exclusion of secondary oil-bearing refinery materials and listed refinery wastes similar to the exclusion currently in place for petroleum cokers. As more performance data become available, a broader exclusion to include other waste materials also seems reasonable.

5.1   Comparison of Available Data from Gasification and Incineration

Publicly available reports on gasification and incineration were accessed to gather data on the environmental performance of each system, the composition of significant byproduct and emissions streams, and also to assess the fate of trace toxic substances within each process. Whenever possible, the fate of the specific toxic constituents in RCRA listed refinery hazardous wastes are addressed based on currently available data. Table 5-1 summarizes the current RCRA listed wastes for the petroleum refining industry. Constituents of concern for which the waste was listed are also provided. These listed hazardous wastes are of specific interest with regard to the proposed extension of the refinery "coker exclusion" to gasification processes as described in the EPA's Notice of Data Availability dated July 15, 1998 (63 FR 38139).

Table 5-1. Summary of RCRA Listed Refinery Wastes

EPA Hazardous Waste Number	Description	Hazardous Constituents for Which Listed
F037	Petroleum refinery primary oil/water/solids separation sludge	Benzene, Benzo(a)pyrene, Chrysene, Lead, Chromium
F038	Petroleum refinery secondary (emulsified) oil/water/solids separation sludge	Benzene, Benzo(a)pyrene, Chrysene, Lead, Chromium
K048	Dissolved air flotation (DAF) float from the petroleum refining industry	Hexavalent chromium, Lead
K049	Slop oil emulsion solids from the petroleum refining industry	Hexavalent chromium, Lead
K050	Heat exchanger bundle cleaning sludge from the petroleum refining industry	Hexavalent chromium
K051	API separator sludge from the petroleum refining industry	Hexavalent chromium, Lead
K052	Tank bottoms (leaded) from the petroleum refining industry	Lead
K169 (1)	Crude oil storage tank sediment from petroleum refining operations	Benzene
K170 (1)	Clarified slurry oil tank sediment and/or in-line filter/separation solids from petroleum refining operations	Benzo(a)pyrene, Dibenz(a,h)anthracene, Benzo(a)anthracene, Benzo(b)fluoranthene, Benzo(k)fluoranthene, 3-methylcholanthrene, 7,12-Dimethylbenz(a)anthracene
K171 (1)	Spent hydrotreating catalyst from petroleum refining operations, including guard beds used to desulfurize feeds to other catalytic reactors.	Benzene, Arsenic
K172 (1)	Spent hydrorefining catalyst from petroleum refining operations, including guard beds used to desulfurize feeds to other catalytic reactors	Benzene, Arsenic

(1) Recently listed wastes as proposed in EPA's final rule 63 FR 42110 dated August 6, 1998 (1).

5.1.1 Gaseous Streams—Major Constituents

The concentration of major and minor gas components for hazardous waste incinerator combustion gases and raw product syngas produced from various feedstocks are presented in Tables 5-2 and 5-3, respectively. These data illustrate the differences in the basic chemical reactions that take place for incineration and gasification. Incinerator combustion gases are composed primarily of nitrogen, carbon dioxide, oxygen and water, with lesser amounts of carbon monoxide, NO_x, SO_2, SO_3, and total unburned hydrocarbons. In contrast, carbon monoxide and hydrogen are the major constituents of the raw syngas, with lesser amounts of carbon dioxide, argon, nitrogen, water, methane, and reduced gas species such as ammonia, hydrogen sulfide, and carbonyl sulfide.

The data in Table 5-3 also illustrate the relatively consistent composition of the raw syngas for various conventional fossil fuels, and mixtures of waste and supplemental fossil fuels, within a given type of gasification technology (e.g., slurry-fed, entrained flow gasifiers). The relative proportions of hydrogen, carbon monoxide and carbon dioxide varies the type of gasification technology because of differences in gasifier design and process conditions. The most significant difference in syngas composition occurs for the reduced sulfur species. As expected, higher concentrations of these species are observed for the higher sulfur fuels such as heavy oil, petroleum coke and the high sulfur coals.

5.1.2 Gaseous Streams—Trace Constituents

Incineration

Trace constituents have been characterized extensively for incineration systems. The trace constituents of concern for hazardous waste incinerator combustion gases have historically been hazardous air pollutants (HAPs), particulate matter, and POHCs in the waste. The list of 189 HAPs in the 1990 CAAA include metals (antimony, arsenic, beryllium, cadmium, chromium, cobalt, lead, manganese, mercury, nickel and selenium), undestroyed POHCs in the waste, PICs and acid gases. PICs include organic substances such as dioxin/furan compounds and PAHs.

Hazardous waste incinerators emit many of the listed HAPs (2). EPA data indicate that metals HAP emissions include antimony, arsenic, beryllium, cadmium, chromium, lead, mercury, nickel and selenium compounds. Organic HAPs emitted include dioxin/furan compounds, benzene, carbon disulfide, chloroform, chloromethane, hexachlorobenzene,

Table 5-2. Typical Composition of Incinerator Combustion Flue Gas

Component	Incinerator Combustion Gas (2,3,4)
H_2, vol %	N
CO, ppmv	10–1500
CO_2, vol %	2–12
O_2, vol %	5–14
Ar, vol %	NR
N_2, vol %	60–90
H_2O, vol %	5–10
NO_x, ppmv	0–4000
SO_2, ppmv	0–4000
SO_3, ppmv	0–100
CH_4, ppmv	N
H_2S, ppmv	N
COS, ppmv	N
NH_3, ppmv	N
THC, ppmv	0.2-36

THC = Total hydrocarbons (excluding methane) expressed as propane.

NR = Not reported.

N = Not present.

methylene chloride, naphthalene, phenol, toluene and xylene. Hydrochloric acid and chlorine gas are present in the combustion gases because of the high chlorine content of many hazardous wastes. Other acid halides (HF and HBr) may also be present depending on the halogen content of the waste feed. Reported trace substance emission data for hazardous waste incinerators are summarized in Table 5-4.

Data for incineration systems indicate that mercury is generally in the vapor form in and downstream of the combustion chamber, including the flue gas cleanup device (2). Thus the level of mercury emissions is a function of the level of mercury in the waste and the use of gas cleanup devices that can control mercury in the vapor form (e.g., carbon injection, wet scrubbers for control of mercury in the soluble HgCl2 form). Other semi-volatile metals (e.g., arsenic, lead, cadmium and selenium) typically vaporize at combustion temperatures and then recondense onto the surface of the fine particulate matter before entering the gas cleanup devices. Emissions of these semi-volatile metals are a function of the waste feed rate and the efficiency of the particulate collection device, particularly the collection efficiency for extremely fine particulate matter. The low-volatile metals such as antimony, barium, chromium, cobalt, manganese and

Table 5-3. Raw Syngas Composition for Various Slagging Gasifier Technologies and Feedstocks

Bed Type: Fuel Form: Fuel Type: Component	Moving Dry feed Ill. #6 Coal (25)	Moving Dry feed Ill. #6 Coal (low pressure) (26)	Entrained Liquid Heavy Oil (5)	Entrained Unknown Pet. Coke (5)	Entrained Dry feed Ill. #5 Coal (25)	Entrained Slurry SUFCO Coal, low S (6)	Entrained Slurry Ill. #6 Coal (6)	Entrained Slurry Pitt. 8 Coal, high S (6)	Entrained Slurry 30/70 Blend of Surrogate RCRA Contaminated Soil with Pitt. 8 Coal (7)	Entrained Slurry 20/80 Blend of Refinery Field Tank Bottoms with SUFCo Coal (7)	Entrained Slurry 25/75 Blend of Municipal Sewage Sludge with Pitt 8 Coal (7)	Entrained Slurry 14/86 Blend of Hydrocarbon Contaminated Soil with Pitt. 8 Coal (7)
H_2, vol %	26.4	52.2	43.32	30.33	26.7	37.6	37.3	37.9	32.3	37.68	35	34.52
CO, vol %	45.8	29.5	45.62	47.72	63.1	41.8	44.0	42.7	34.6	39.45	38.5	48.36
CO_2, vol %	2.9	5.6	8.17	17.88	1.5	19.8	16.9	17.3	26.3	21.21	23.5	15.64
O_2, vol %	N	N	N	N	N	N	N	N	N	N	N	N
Ar, vol %	NR	NR	1.00	0.83	NR	0.08	0.08	0.07	0.1	0.08	0.1	0.08
N_2, vol %	3.3	1.5	0.53	1.27	5.2	0.69	1.1	1.41	5.8	1.32	1.9	0.18
H_2O, vol %	16.3	5.1	0.27	0.12	2	NR	NR	NR	NR	NR	NR	NR
NO_x, ppmv	N	N	N	N	N	N	N	N	N	N	N	N
SO_2, ppmv	N	N	N	N	N	N	N	N	N	N	N	N
SO_3, ppmv	N	N	N	N	N	N	N	N	N	N	N	N
CH_4, ppmv	38,000	44,000	3500	100	300	2400	1570	1930	60	300	420	NR
H_2S, ppmv	10,000	9000	7100	10,760	13,000	1260	9570	7590	2070	NR	NR	NR
COS, ppmv	1000	400	0.00 [a]	20	1000	23.2	153	176	140	NR	NR	NR
NH_3, ppmv	2000	5000	0.00 [a]	0.00 [a]	200	2.3	0.58 [a]	0.62 [a]	NR	NR	NR	NR
THC, ppmv	2000	3000	NR	NR	NR	NR	NR	NR	27	NR	NR	NR

[a] Measured after particulate scrubbing and gas cooling (i.e., after ammonia removal).

THC = Total hydrocarbons (excluding methane) expressed as methane.

NR = Not reported.

N = Not present.

Table 5-4. Reported Trace Substance Emissions
from Hazardous Waste Incineration

Substance	Median	Range	No. of Measurements
Particulate Matter, mg/dscm	32	0.0072 - 12,800	632
Metals, µg/dscm			
Mercury	9.5	0.04 - 2400	177
Cadmium	6.9	0.022 - 1890	256
Lead	92	0.44 - 53300	241
Selenium	1.3	0.31 - 47	72
Antimony	7.3	0.12 - 156000	164
Arsenic	4.4	0.079 - 1180	253
Barium	30	0.27 - 1050	99
Beryllium	0.25	0.0079 - 57	213
Chromium	25	0.094 - 923	272
Nickel	35	0.22 - 2050	155
Silver	2.5	0.015 - 1320	137
Thallium	5.2	0.13 - 361	130
Organics,			
Dioxin/Furans	0.25 ng/dscm (TEQ basis)	0.0047 - 77 ng/dscm (TEQ basis)	141
Semi-volatile Organics (46 compounds detected)	-	21 - 330000000 ng/dscm	3 - 64
Volatile Organics (59 compounds detected)	-	3.2 - 7050000 ng/dscm	3 - 141
Inorganics, ppmv			
HCl	5.6	0.038 - 949	472
HF	0.24	0.063 - 0.54	9

Source: Reference 8.

nickel are less likely to vaporize in the combustion process, and thus they partition to the bottom ash in the combustion chamber and to the large, easy-to-control particles in the combustion gas. Thus, emissions of low-volatile metals are more strongly related to the operation of the particulate collection device.

EPA's database for hazardous waste incinerators includes data for 46 SVOCs and 59 VOCs detected in the combustion gases over a wide range of concentrations (8). Dioxin/furan compounds are also often detected in the combustion gases from hazardous waste incinerators. The volatile organic compounds tend to be detected more often and at higher concentrations than the SVOCs. These PICs can result from:

- Incomplete destruction of POHCs in the waste;

- New compounds created in the combustion zone and downstream as the result of reactions with other compounds or compound fragments;

- Compounds present in the waste feed but not identified as a POHC; or

- Compounds from other sources such as ambient air used for combustion.

Gasification

Similar data for gasifier product syngas and turbine/HRSG stack emissions are much more limited. The most comprehensive trace substance characterization tests were conducted for single-stage entrained bed gasifiers at the CWCGP, SCGP-1, and more recently, for a two-stage entrained bed LGTI gasifier (6,28,19). These studies were conducted during the gasification of various coal feedstocks and did not include gasification of hazardous wastes. As mentioned earlier in this section, waste gasification tests were conducted on a single stage, entrained bed gasification process as part of the EPA's Superfund Innovative Technology Evaluation (SITE) Program in 1994 (7). Less comprehensive test data are also available for refinery gasification operations (29,30,31,32) and waste gasification processes (33,34,35,36,37).

The SITE program tests were designed to evaluate the ability of a single-stage, entrained bed gasification process to treat hazardous waste material (contaminated soil) containing both organic compounds and inorganic heavy metals. The POHC - chlorobenzene - was spiked into the waste feed to determine DRE for the waste gasification tests. In addition, barium and lead were spiked into the waste soil feed to ensure that the levels in the waste were sufficient for the waste to fail the RCRA TCLP characteristic test. These test data provide some valuable information on the performance of waste gasification systems. However, the tests did not address gas turbine/HRSG or incinerator emissions because the syngas and other internal gases from the pilot-scale test facility were flared. Information on the partitioning of the spiked metals (barium and lead) was provided, but the concentrations of metals in the gas streams were not determined.

Combustion gas from the small incinerator treatment system, commonly used to treat residual gaseous streams in a gasification plant, is also an air emission source that must be considered. However, the volumetric flow rate from the incinerator stack is typically less than 1% of the volumetric flow rate from the turbine exhaust. Table 5-5 provides a comparison of total air emissions from the CWCGP and LGTI coal gasification units (turbine plus incinerator stack).

Criteria pollutant emissions from the CWCGP and LGTI gasification tests were very similar and indicative of the very low emission rates associated with gasification processes. Particulate emissions, expressed on a concentration basis, were 5.5 mg/dscm for CWCGP and

Table 5-5. Comparison of Total Air Emissions (Turbine and Incinerator Stack) from Coal Gasification Systems

Substance	Coal Gasification (lb/trillion Btu input)	
	CWCGP (Illinois 6, High S Bit)	LGTI (Powder River Coal, Low S Subbit)
Criteria Pollutants		
Particulate	0.0086 lb/10^6 Btu	0.0091 lb/10^6 Btu
Sulfur dioxide	0.075 lb/10^6 Btu	0.12 lb/10^6 Btu
Nitrogen oxides	0.09 lb/10^6 Btu	0.26 lb/10^6 Btu
Ionic Species		
Ammonia (as N)	<230	440
Chloride	NS	740
Fluoride	<13	38
Cyanide	NS [a]	0.08
Metals		
Antimony	1.4	4
Arsenic	9.5	2.1
Barium	37	NS
Beryllium	<6	0.09
Cadmium	<7	2.9
Chromium	185	2.7
Cobalt	5.7	0.57
Lead	162	2.9
Manganese	18	3.1
Mercury	NS	1.7
Nickel	94	3.9
Selenium	15	2.9
Silver	<7	NS
Aldehydes		
Acetaldehyde	NS	1.8
Benzaldehyde	NS	2.9
Formaldehyde	NS	17
Volatile Organic Compounds		
Benzene	<5	4.4
Carbon disulfide	<750	46
Toluene	NS	0.033
PAHs/SVOCs		
2-Methylnaphthalene	ND [b]	0.36
Acenaphthylene	ND	0.026
Benzo(a)anthracene	ND	0.0023
Benzo(e)pyrene	ND	0.0056
Benzo(ghi)perylene	ND	0.0096
Naphthalene	ND	0.4

[a] NS = Not sampled.

[b] ND = Not detected. Detection limits not reported. None of the 80 SVOCs tested were detected in either the turbine/HRSG or incinerator stack.

5.1 mg/dscm for LGTI. These concentrations are two orders of magnitude lower than the current RCRA particulate emission standard for hazardous waste incinerators (180 mg/dscm) and one order of magnitude lower than the recently proposed MACT standard for new and existing hazardous waste incinerators (34 mg/dscm). Particulate matter concentrations less than 10 mg/dscm in the gas turbine emissions have been reported for a gasification system using heavy refinery residual feedstocks such as vacuum visbroken residue, vacuum residue, and asphalt (32).

Stack emissions of arsenic, chromium, cobalt, lead, manganese, nickel and selenium were higher for the CWCGP process compared to the LGTI process. Differences in trace metals content of the two coal feedstocks may partially explain the observed difference. The concentration of these metals in the Illinois 6 coal feedstock during the CWCGP test was typically a factor of 2 to 13 higher than the concentrations in the Powder River coal used at LGTI.

None of the volatile or semi-volatile organic compounds (including PAHs) were detected in the turbine stack or incinerator stack during the CWCGP tests. Volatile organic compounds detected during the LGTI test included benzene, toluene and carbon disulfide (all on the order of parts per billion in the combustion gases). Semi-volatile organic compounds detected were primarily PAH compounds and were typically detected at sub parts per billion concentrations. Tests for the SCGP-1 process indicated that PAHs and phenolic compounds were not detected in the raw syngas at a detection limit of approximately 1 ppbv. The total concentration of other non-methane hydrocarbons varied from 0.5 to 90 ppbw in the raw syngas.

Results from the waste gasification tests conducted as part of the SITE program also indicated the presence of selected volatile and semi-volatile organic compounds in the raw syngas, clean syngas, flash gases, and sulfur removal acid gases at sub parts per billion levels. Chlorobenzene was spiked into the waste feed stream during these tests to determine the DRE for the POHC chlorobenzene, resulting in a high chlorine content waste feed. Carbon disulfide, benzene, toluene, naphthalene, naphthalene derivatives, and acenaphthalene were measured in the gas streams at parts per billion levels. The POHC chlorobenzene was also detected, although the tests demonstrated DREs greater than 99.9956 for chlorobenzene. Since the syngas was not combusted in a turbine, and the flash gas and acid gas streams were not incinerated at this pilot-scale test facility, the levels of these compounds in the resulting combustion gas can not be determined. However, one would expect further destruction of these compounds as a result of combustion in the turbine and incinerator systems.

Additional testing of the same gasification system used in the SITE program was conducted between 1988-1991 while feeding residual materials such as petroleum tank bottoms, municipal sewage sludge, and hydrocarbon-contaminated soil. Results indicated that no organic compound heavier than methane was present in the raw syngas at concentrations greater than 1 ppmv, consistent with the requirements of the comparable fuels exclusion (7).

Recent test data have been reported for gasification of highly chlorinated feedstocks in an entrained bed system. Feedstock included chlorinated heavies from the production of dichloropropane and dichloroethane (36). Syngas measurements from these tests indicated none of the chlorinated VOCs (except chloroform) were detected in the syngas at a detection limit of approximately 1 ppbv. Benzene, toluene, ethylbenzene, chloroform and xylene were detected at ppbv levels.

Tests conducted for a pilot-scale fixed bed slagging gasifier also indicate extremely high destruction efficiencies for hexachlorobenzene, a surrogate compound used to simulate waste feeds containing PCBs (34). DREs for hexachlorobenzene were greater than 99.9999% and hexachlorobenzene was not detected in the in the product syngas.

5.1.3 Polychlorinated Dibenzo Dioxins and Furans

Dioxin and furan compound emissions from incineration systems have been studied extensively over the past 10 years (9-18). Formation of these compounds has been shown to occur in the combustion process and in the downstream combustion gas treatment processes. A detailed discussion of the formation mechanism is beyond the scope of this paper, however, the major mechanisms of formation can be summarized as follows:

- In-furnace Formation—Hydrocarbon precursors react with chlorinated compounds or complex organic molecules in the combustion process to create dioxin/furan compounds; and/or

- Post-combustion Formation—Gas-phase condensation of dioxin precursors onto fly ash in the cooler post-combustion regions and subsequent formation on the particulate surface via catalytic reactions.

The latter is thought to be the predominant formation mechanism of incinerator emissions. The post combustion mechanism involves the low temperature formation within the post-combustion zone in the presence of free chlorine (Cl_2), unburned carbon or precursors, and copper catalyst species in the fly ash (13). It has been shown that free chlorine is produced in the post combustion processes of waste incinerators via the Decon reaction in which HCl is converted to Cl_2 on the fly ash surface in the presence of oxygen and copper catalysts. Free chlorine then chlorinates the aromatic ring structures of precursors through substitution reactions. HCl has been shown to be a relatively ineffective chlorinating agent in the production of dioxin/furan compounds. The optimum temperature window for formation of dioxin/furan compounds for incineration systems has been shown to be 450°F to 650°F (decreasing formation with decreasing temperature) (2). Thus, the recent final MACT standard for hazardous waste incineration systems have specifically addressed the issue of gas temperatures at the inlet to the APCD (air pollution control device) by setting a maximum allowable temperature of 400°F for existing hazardous waste incinerators (27).

Dioxin and furan compounds are not expected to be present in the syngas from gasification systems for two reasons. First, the high temperatures in the gasification process effectively destroy any dioxin/furan compounds or precursors in the feed. Secondly, the lack of oxygen in the reduced gas environment would preclude the formation of the free chlorine from HCl via the Decon reaction, thus limiting chlorination of any dioxin/furan precursors in the syngas. If the syngas is combusted in a gas turbine, one would not expect formation of dioxin/furan compounds because very little of the particulate matter required for post combustion formation is present in the clean syngas or in the downstream combustion gases entering the HRSG system. In addition, the temperature profiles in the combustion turbine where oxygen is present are not in the favorable range for the Decon reaction (660 - 1,290 °F) (20), so production of free chlorine from HCl will be limited.

Measurements of dioxin/furan compounds in gasification systems confirm these expectations. Dioxins were not measured as part of the CWCGP or LGTI tests; however, they were measured in the gas streams as part of the SITE program tests. Measured concentrations of PCDDs and PCDFs in the gas streams (i.e., raw syngas, clean syngas, sulfur removal acid gas, and flash gas) were all comparable to the blanks, indicating that these species, if present, were at concentrations less than or equal to the method detection limits (parts per quadrillion, ~ 0.01 ng/Nm3). The GTC has reported dioxin/furan data from a high-temperature, moving-bed waste gasification facility that are an order of magnitude below the recently finalized MACT standard for new and existing hazardous waste incinerators (MACT = 0.2 ng/Nm^3 TEQ, measured = 0.02 ng/Nm^3 TEQ) (21, 37). Measurement results from a waste gasification facility in Germany have also shown extremely low levels of PCDD/PCDF compounds in the clean product syngas (less than 0.002 ng/Nm3 TEQ (35).

PCDD/PCDF data in syngas produced from pilot-scale gasification of highly chlorinated feedstocks such as hexachlorobenzene and chlorinated heavies from the manufacture of 1,2-Dichloropropane (DCP) and 1,2-Dichloroethane (DCE) have also been reported. In the first test program, hexachlorobenzene was gasified in a fixed bed gasifier with petroleum coke to simulate destruction of PCB compounds. Syngas measurements indicated PCDD/PCDF concentrations ranging from 0.001 to 0.03 ng/Nm^3 TEQ (34). In the second test program, chlorinated heavies from the manufacture of DCP and DCE were gasified in an entrained bed gasifier to demonstrate a process for syngas production and HCl byproduct recovery. PCDD/PCDF concentrations in the syngas were very near the method detection limits, on the order of 0.001 ng/Nm^3 TEQ (36).

5.1.4 Fate of Trace Metals and Halides in Gasification Systems

Since hazardous wastes were not included in the feedstocks for the CWCGP and LGTI tests, specific conclusions regarding the level of trace constituents emitted from the gas turbine stack and incinerator stack during gasification of hazardous wastes can not be directly drawn. However, the data from these test programs and the SITE program tests do provide valuable insight on the general fate of toxic substances (particularly metals) in gasification systems. The material balance results from the CWCGP and LGTI test programs are summarized in Tables 5-6 and 5-7, respectively. The partitioning of selected volatile/semi-volatile and non-volatile elements among the various discharge streams is shown graphically in Figures 5-1 and 5-2, respectively.

Volatile trace elements such as mercury, chloride and fluoride are vaporized almost completely during gasification and are carried downstream in the process. Only small portions, if any, are retained in the slag as shown in Figure 5-1. Chloride and fluoride are typically removed during the gas cooling and scrubbing operations, and ultimately partition to the water systems Some may also remain in the product syngas and exit in the turbine exhaust. Data from the SITE program tests showed greater than 99% removal of hydrogen chloride from the syngas. A gasification processes using highly chlorinated feedstocks is currently being tested to develop a process for removal and recovery of the HCl as an anhydrous or aqueous product (36).

Semi-volatile trace elements such as arsenic, cadmium, lead and selenium partition partially into the slag, but can also be present in the vapor phase throughout the process. Data from the LGTI tests indicate lead and cadmium partitioned slightly into the water system. These substances may also volatilize and recondense on the fine particulate matter as the syngas cools. Data from the SITE program indicated that the concentration of lead in the clarifier solids (fine

Table 5-6. Elemental Flow Rates Around the CWCGP
Gasification Process, Illinois 6 Coal Test (lb/hr)

| Substance | Input | Output | | | | | Percent Closure |
	Coal Feed	Slag	Incinerator Stack [a]	Gas Turbine Stack	Sulfur Byproduct	Aqueous Discharges to Evap. Pond	
Antimony	<0.083	0.078	<0.00011	<0.0086	IS [b]	IS	>94
Chloride	125	0.0036	<0.00061	NR [c]	IS	64	50
Fluoride	13	0.00032	<0.00020	<0.014	IS	0.8	6
Arsenic	0.55	0.24	0.00026	0.01	IS	IS	45
Barium	4.3	2.57	0.00056	0.04	IS	IS	60
Beryllium	<0.04	0.04	<0.000086	<0.0062	IS	IS	100
Cadmium	0.054	0.04	0.00016	<0.0078	IS	IS	74
Chromium	1.75	3.7	0.053	0.15	IS	IS	220
Cobalt	0.34	0.41	0.00033	0.0058	IS	IS	120
Copper							
Lead	1.5	0.96	0.00054	0.18	IS	IS	76
Manganese	2.8	3.3	0.008	0.013	IS	IS	118
Mercury	0.005	0.0009	<0.0017	NS	IS	IS	20-50
Nickel	1	1.45	0.036	0.066	IS	IS	155
Selenium	<0.025 [d]	0.067	<0.000065	0.016	IS	IS	>330
Aluminum	916	861	IS	IS	IS	IS	94
Calcium	358	347	IS	IS	IS	IS	97
Iron	1500	1230	IS	IS	IS	IS	82
Magnesium	48	46	IS	IS	IS	IS	96
Potassium	158	156	IS	IS	IS	IS	99
Sulfur	2417	168	6.2	35	2208 (diff.)	IS	100

Source: Reference 6.
[a] Vapor-phase metals only.
[b] IS = Insignificant. <1% of the total input.
[c] NR = Not reported.
[d] Coal analysis is suspect.

Table 5-7. Elemental Flow Rates Around the LGTI Gasification Process (lb/hr)

Substance	Input Coal Feed	Output [a] Slag	Incinerator Stack	Gas Turbine Stack	Sulfur Byproduct	SWS Discharge	Percent Closure
Antimony	0.016	0.011	0.00009	<0.011	IS [b]	IS	69-138
Chloride	5.3	0.83	0.09	2.0	IS	0.07	57
Fluoride	10	2.0	0.0012	0.1	IS	0.15	22
Arsenic	0.13	0.059	0.000029	0.0056	IS	IS	50
Beryllium	0.037	0.034	0.0000027	<0.00025	IS	IS	95
Cadmium	0.014	0.002	0.000083	0.0077	IS	0.0004	78
Chromium	0.64	0.76	0.00016	0.0071	IS	IS	120
Cobalt	0.26	0.26	0.000016	0.0015	IS	IS	101
Copper	1.6	1.5	0.00011	0.04	IS	IS	100
Lead	0.18	0.03	0.000093	0.0076	IS	0.027	37
Manganese	1.3	1.3	0.00041	0.0080	IS	IS	99
Mercury	0.015	0.00020	0.0012	0.0034	IS	IS	33
Molybdenum	0.074	0.075	0.00022	0.018	IS	IS	134
Nickel	0.21	0.38	0.00022	0.010	IS	IS	187
Selenium	0.45	0.013	0.0000098	0.0080	IS	IS	33
Aluminum	850	900	<0.003	0.2	IS	IS	107
Calcium	1400	1600	<0.004	0.6	IS	IS	111
Iron	330	370	0.009	0.4	IS	IS	113
Magnesium	300	330	<0.001	0.08	IS	IS	109
Potassium	28	27	<0.01	0.9	IS	IS	98
Sulfur	380	3.0	170	38	240	IS	119

Source: Reference 19.

[a] Scrubber solids and water are recycled to the gasifier in this process, so they do not represent and output.

[b] IS = Insignificant. <1% of the total input.

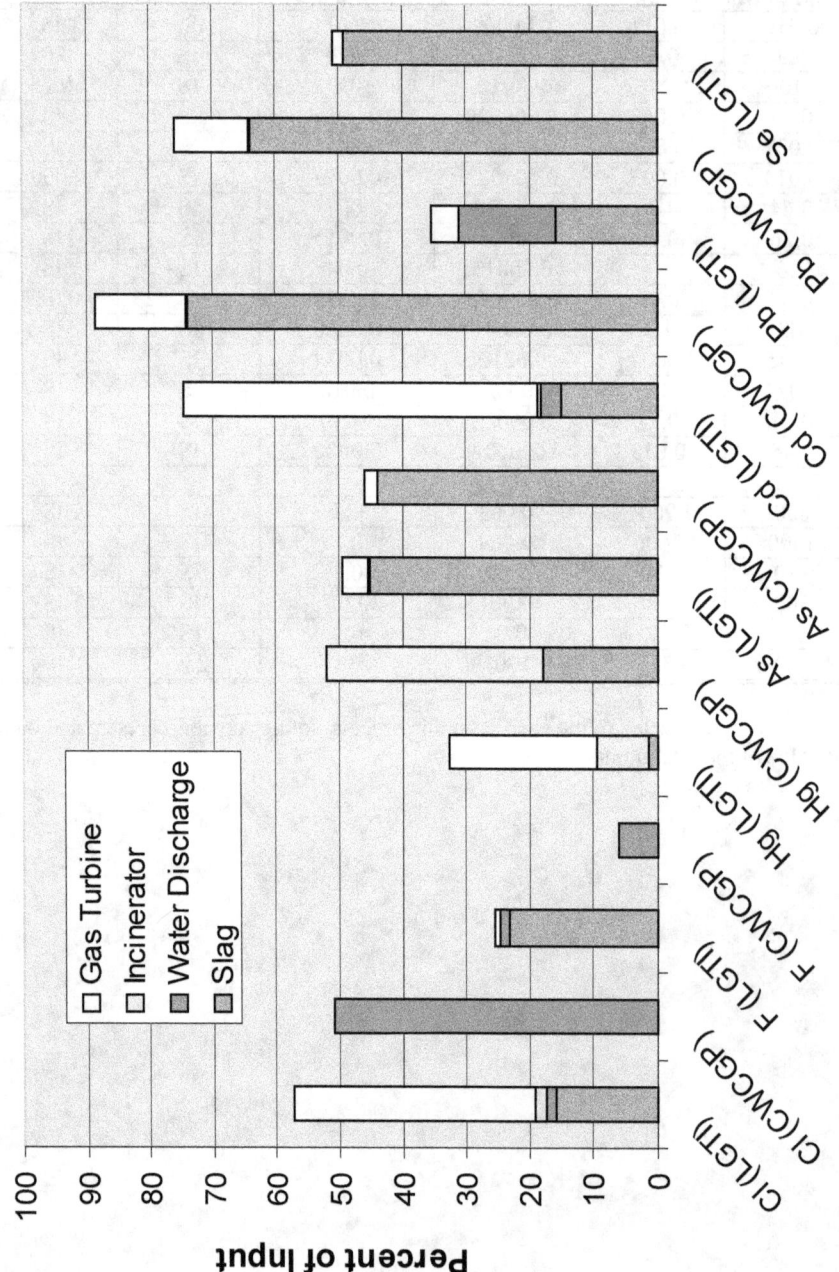

Figure 5-1. Partitioning of Volatile Trace Substances in Gasification Systems

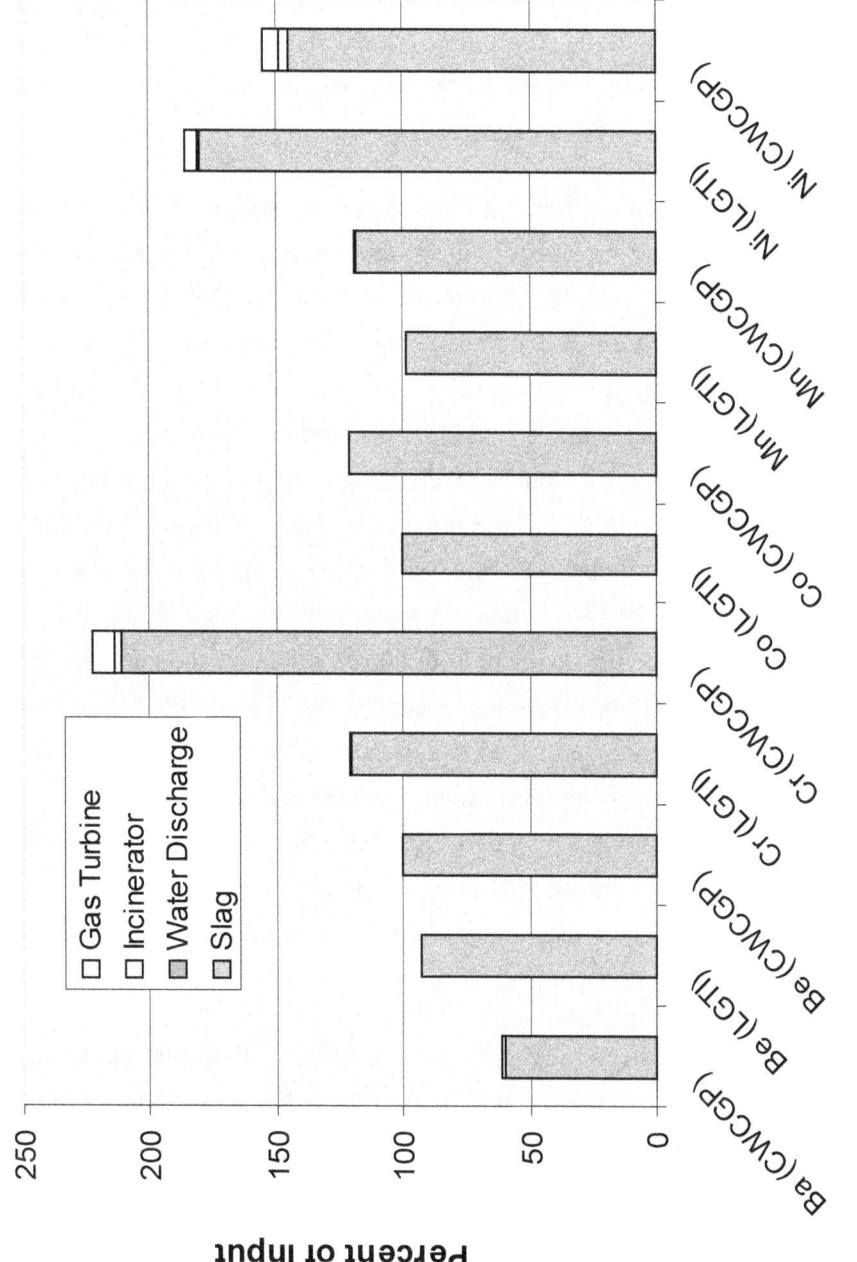

Figure 5-2. Partitioning of Non-Volatile Trace Substances in Gasification Systems

particulate matter collected from the syngas during gas scrubbing) was substantially enriched compared to the coarse slag (55,000 ppmw vs. 391 ppmw).

The mass balance closures for the volatile and semi-volatile elements were substantially less than 100% during both the LGTI and CWCGP test programs, so the fate of these elements remains uncertain. Results from the SITE program indicated only about 30% of the total lead input with the feedstocks was accounted for in the slag materials and the clarifier solids. Lead concentrations in the gas streams were not reported.

Mass balance results from the SCGP-1 test program during gasification of various coals are similar (28). The trace element content of the product syngas and the acid gas stream to the sulfur recovery unit was determined rather than a turbine exhaust and tail-gas incinerator stack stream, so these data are not included in Figures 5-1 and 5-2. Mass balance closures for the volatile and semi-volatile trace elements were also substantially less than 100% for the SCGP-1 tests. The low recoveries were shown to be evidence of retention of volatile trace elements with the process equipment. Volatile trace elements were not detected in the clean product syngas or the acid gas, with the exception of lead (clean syngas) and selenium (acid gas) which were present at less than 1% of the total inlet feed rate to the gasifier. Analyses of precipitated deposits from the packing material within the syngas washing step showed the solids to be highly enriched in elements such as mercury, arsenic, lead, nickel, selenium and zinc. In addition, analyses of the SCGP-1 Sulfionol-based solvent used in the acid gas removal system after several thousand hours of operation confirmed that most trace elements were removed from the syngas before it entered the acid gas removal system. Additional tests during gasification of petroleum coke in the SCGP-1 process provided data on the fate of nickel and vanadium at concentrations about two orders of magnitude higher than those typically found in coal. Results showed that concentrations of nickel and vanadium in the clean product syngas were below the limits of detection (7 and 2 ppbw, respectively)(9).

There is also some evidence to suggest that some of the volatile and semi-volatile trace elements may accumulate in the amine-based solvents used in the sulfur removal systems at gasification facilities. These solvents are periodically regenerated to prevent the buildup of heat stable salts. These salts may retain metallic elements in solution by chelation, or the high sulfide levels may force the precipitation of metals sulfides. In the case of mercury, LGTI test data showed that concentrations in the tail-gas incinerator stack ($28 \ \mu g/Nm^3$) were significantly higher than concentrations measured in the turbine exhaust ($0.71 \ \mu g/Nm^3$). As discussed in the LGTI report, a possible explanation is the formation of mercuric sulfide in the syngas which would be removed from the gas by the amine-based solvent (MDEA) in the Selectamine absorber. When the solvent is regenerated, the mercury would be desorbed into the acid gas stream going to the sulfur removal unit. The volatile mercury would pass through the sulfur removal system and exit in the tail gas that was routed to the small tail gas incinerator.

Trace substances typically considered non-volatile include barium, beryllium, chromium, cobalt, manganese and nickel. In most cases, these substances partitioned almost entirely to the slag as shown in Figure 5-2. During the SITE test program, barium was spiked into the waste feed stream to determine its fate in the system. Over 90% of the barium input to the gasifier was accounted for the in the slag materials and 2% was found in the clarifier solids which is consistent with the behavior of a non-volatile substance. Data from both the LGTI and CWCGP test programs show higher than expected concentrations of chromium and nickel in the slag when compared to the amount input with the fuels. The most likely source is the refractory material used to line gasifier reactor. Chromium and nickel were also found in the turbine exhaust during both test programs.

Gasification of petroleum-based feedstock such as petroleum coke and heavy refinery residues that contain high concentrations of nickel and vanadium have also shown that these elements are effectively captured and concentrated in the slag material (30,31,38). Recent test data from a two-stage entrained bed feeding petroleum coke have shown 80% of the nickel and 99% of the vanadium fed to the gasifier were captured in the slag (38). Nickel and vanadium were present in the coke feedstock at concentrations as high as 300 and 1500 ppmw, respectively. They were not detected in the liquid or gas streams resulting from gasification.

5.1.5 Solid Byproducts

For hazardous waste incinerators, RCRA requirements mandate that any ash from combustion chamber and downstream gas cleanup devices is also considered a hazardous waste. The principal contaminants are heavy metals primarily in the form of metal oxides, and undestroyed organic material. Test data suggest that very small amounts of residual organic compounds remain in incinerator ash and control device residuals. When organic compounds were detected, they tended to be toluene, phenol and naphthalene at concentrations less than 30 parts per billion (22,23).

Analysis of the slag material produced from various coal gasification processes has consistently shown the slag to be a nonhazardous waste according to RCRA definitions. Trace metals tend to concentrate in the slag; however, the glassy slag matrix effectively immobilizes the metals eliminating or reducing their leachability. None of the slags produced during the gasification of 4 coals at the CWCGP demonstration facility exhibited any of the RCRA waste characteristics and would have been ruled nonhazardous (6).

Test data from the entrained-bed gasification system at the El Dorado refinery during gasification of acid soluble oils and phenolic residue have shown that the slag and the fine particulate matter removed from the raw syngas both passed the TCLP analyses (29). The fine particulate matter is currently recycled to the gasifier to recover the energy value of the unreacted carbon in the solids.

Recent test data for a two-stage, entrained bed gasifier feeding petroleum coke have shown that the resulting slag passed TCLP leachate tests and was classified as nonhazardous (38).

During the waste gasification tests conducted for the SITE program, leachability characteristics of the coarse slag, fine slag and clarifier solids were tested using TCLP and WET-STIC methods. The test slurry feed was spiked with barium nitrate and lead nitrate to create a surrogate RCRA-hazardous waste feed. Results from the SITE tests are summarized in Table 5-8. As discussed above, lead was highly enriched in the fine particulate matter removed from the syngas as evidenced by the high concentration in the clarifier solids. Although the clarifier solids comprised only about 1.6% of the total solid residuals, they contained 71% of the lead measured in all of the solid residual streams. In contrast, barium partitioned to the solid residual streams in approximate proportion to the mass flow of each stream.

TCLP and WET-STLC results for the slurry feed were above the regulatory limits for lead and below the limits for barium. The test results also showed that the waste gasification process can produce a major solid residual (coarse slag) with TCLP measurements below regulatory limits for both lead and barium. TCLP results for the fine slag and clarifier solids were also below the regulatory limits for barium, but exceeded the limits for lead. The WET-STLC measurements for lead exceeded the regulatory limits for all of the solid residual streams. However, the gasification process did demonstrate significant improvements in reducing the mobility of lead. It is important to emphasize that residual fines collected from the raw syngas represent only a small percentage of material compared to the slag and that these solids are often recycled to the gasifier to recover their energy value.

Table 5-8. SITE Program Test Results for Solid Residuals from Waste Gasification

	Flow Rate (lb/hr)	Lead	Barium
Slurry	2,216	880 ppmw	2,700 ppmw
Coarse Slag	273	329 ppmw	11,500 ppmw
Fine Slag	157	491 ppmw	15,300 ppmw
Clarifier Solids	6.8	55,000 ppmw	21,000 ppmw
TCLP (mg/L)			
Slurry	-	8.3	0.1
Coarse Slag	-	4.5	0.6
Fine Slag	-	14.9	1.75
Clarifier Solids	-	953	2.7
Regulatory Limit	-	**5**	**100**
WET-STLC (mg/l),			
Slurry	-	56	<5.5
Coarse Slag	-	9.8	<5
Fine Slag	-	43	9.3
Clarifier Solids	-	1,167	38.4
Regulatory Limit	-	**5**	**100**

Source: Reference 7.

Additional waste gasification tests have been conducted on petroleum production tank bottoms, municipal sewage sludge and hydrocarbon-contaminated soil at the same pilot-scale gasification facility used for the SITE program (7). The gasification of petroleum tank bottoms was conducted as part of a study for the California Department of Health Services (contract 88-T0339). The tank bottoms were contaminated with 3,000 ppmw benzene, toluene, ethylbenzene and xylene. The tests successfully converted the RCRA-exempt, low-Btu hazardous waste to syngas and produced nonhazardous byproducts and effluents.

Test results from the gasification of municipal sewage sludge also showed that the volatile heavy metals tended to partition to the clarifier solids. Lead was present in the slurry feed at concentrations of about 190 ppmw and nearly 86 weight percent of the recovered lead was found in the clarifier solids. The coarse slag and fine slag streams did not exceed TCLP limits for any metal. The clarifier solids, which represented only 3% of the total residual solids, exceeded TCLP limits for cadmium and lead. Again, in a full-scale gasification unit these clarifier solids would typically be recycled to the gasifier or processed further for reclamation of metals.

Similar results were obtained for the gasification of hydrocarbon-contaminated soil that contained 4 weight percent heavy vacuum gas oil from a refinery. The coarse and fine slag were nonhazardous based on Federal and California standards. The low-volume clarifier solids were above only the California WET-STLC limits for arsenic and lead.

Other low volume residuals from the gasification process may be produced in the sulfur removal and recovery operations and include solids such as solvent filter cakes and spent catalyst material from the sulfur recovery process. These residuals are not unique to the gasification process since they are typically generated in other common applications of these technologies in the natural gas and refinery industries. However, these solids may contains metal sulfides and other metal compounds removed from raw syngas.

5.1.6 Liquid Byproduct and Wastewater Streams

Gasification technologies typically produce only two liquid discharge/byproduct streams: treated water discharge, and liquid sulfur byproduct (sulfuric acid byproducts can also be produced, but this is less common). Liquid sulfur produced from cleanup of gasification syngas is typically over 99% pure. Trace levels (parts per million) of some metals (chromium, mercury and selenium) have been measured in the sulfur byproduct produced at the LGTI and CWCGP coal gasification facilities. Internal residual water streams in the gasification process are typically recycled to the process (e.g., syngas scrubber blowdown that is flashed and recycled to slurry preparation), or treated using common industrial treatment processes such as steam stripping (e.g., syngas cooling condensate processed in a sour water stripper). These water streams may contain trace levels of volatile and semi-volatile organic compounds as indicated by the data collected during the SITE program waste gasification tests. Compounds such as benzene, acetone, carbon disulfide, naphthalene derivatives, and fluorene were detected in the syngas cooling condensate and scrubber water recycled to the process. No concentrations of PCDDs or PCDFs were found above the method detection limit of 10 nanograms per liter (ng/l). Inorganic substances such as ammonia, chloride, and trace metals such as lead were also detected.

During the tests conducted at the LGTI facility, the particulate scrubber water was recycled to the feed preparation area via a char slurry and the syngas cooling condensate was treated a sour water stripper (SWS). Most of the treated water was subsequently recycled to the process and a portion of the treated water was discharged to a permitted outfall. The treated discharge from the SWS was analyzed and contained the following: trace metals at parts per billion levels (μg/L); ammonia, chloride, cyanide, fluoride, formate and thiocyanate at parts per million levels (mg/L), volatile organic compounds at part per billion levels (1,4-Bromofouorobenzene, acetone); and SVOCs (2,4,6-Tribromophenol, 2-Fluorobiphenyl, 2-Fluorophenol, 4-Methylphenol/3Methylphenol, Benzoic acid, Fluoranthene, Phenol, and Pyrene) at concentrations in the range of 0.5 μg/L to 400 μg/L. In most refinery applications, excess pre-treated water from the sour water stripper that is not recycled to the process undergoes final treatment in conventional waste water treatment systems. The standard system at many refineries is flocculation/sedimentation followed by biological treatment (MPG MARS paper, Shell brochure). Special tests were conducted during the gasification of phenolic residue and petroleum coke at the El Dorado refinery to determine if phenol was present in the gasification process water blowdown sent to the refinery WWT system. Phenol was not detected (29).

In hazardous waste incineration systems, aqueous residual streams typically contain entrained particulate matter, absorbed acid gases (usually HCl) and small amounts of organic contaminants. The major internal wastewater stream is the scrubber water used to remove particulate and/or acid gases from the combustion gases. This stream can be highly acidic and may require neutralization before physical/chemical treatment to remove dissolved and suspended metals. In the wet particulate removal processes substantial amounts of metals can leach from the ash, particularly under the acidic (low pH) conditions caused by HCl removal in the scrubber. In contrast, the pH of the quench and scrubber water from gasification systems is typically neutral because the ammonia removed from the syngas during the operations tends to neutralize the HCl in the scrubber water. Settled solids from incinerator scrubber waters are typically disposed as a hazardous waste, and a portion of the treated water may be recycled. In gasification systems, the entrained solids in the scrubber water are typically recycled to the process along with the water after dissolved gases have been removed in a flash vessel. Results from a ten-incinerator test program indicated the presence of 9 volatile and 5 semi-volatile organic compounds in the scrubber water. Semi-volatiles ranged from 0 to 100 µg/L while volatile compounds were generally found at much higher concentrations (0 to 32 mg/L) (23).

5.2   Data Gaps

What is lacking from the published literature on *waste* gasification is a comprehensive assessment of hazardous substances throughout the system. On coal, the most recent assessment of hazardous substances in a gasification system was obtained in 1995 during Phase 2 of the DOE's Comprehensive Assessment of Air Toxics Program. Radian Corporation conducted a thorough multi-phase sampling and analytical test program at a coal-fired IGCC facility located in Plaquemine, Louisiana and operated by the Louisiana Gasification Technology Inc.(LGTI).

The test involved the chemical characterization of over twenty process streams from the raw coal feed to the gas turbine/HRSG emissions. Characterization of most process streams included analyses for ionic species (halides, ammonia, cyanide), metals (including mercury), aldehydes, PAHs/SVOCs, and volatile organic compounds. Polychlorinated dibenzodioxins and dibenzofurans (PCDD/PCDFs) were not included in the testing. Figure 5-3 illustrates the sampling locations in a process flow diagram and Table 5-9 lists the test parameters for each stream. Among the findings of this assessment was the significant reduction in combined emissions compared to a well-controlled, pulverized coal-fired, steam-electric generating station.

A sampling and analytical program of this magnitude may not be necessary to support the extension of the petroleum coker exclusion to include gasification. However, the results from a well planned series of demonstration tests under baseline and waste feed conditions could fill the existing data gaps. This would provide the EPA and industry with the information necessary to support a broader exclusion to include other RCRA wastes and enable the expansion of this technology. Assuming that gasification performs to the same level of environmental standards it has set for the gasification of coal for electric power production, this could achieve a net positive effect on the environment.

5.3   Status of Sampling and Analytical Methods for Gasification Processes

Another important issue raised in the LGTI report is the availability of suitable sampling and analytical methods for synthesis gas streams. This issue focuses primarily on vapor-phase metals in syngas, but it is important to consider the effects of synthesis gas components, typically in a reduced and reactive form, on the collection mechanisms commonly employed by flue gas emissions testing. Depending on the sampling location, gas moisture, which may condense in sampling lines or headers, is another important consideration that must be addressed when planning the sampling approach.

While many conventional flue gas-testing methods have been applied to syngas, none have ever been validated for that purpose. Therefore, it is strongly recommended that any proposal to provide characterizations of synthesis gas or other gasification process streams be accompanied by a thorough data quality control and quality assurance plan that incorporates measures to assess the performance of the sampling and analytical methods applied.

Table 5-10 summarizes the effectiveness of syngas sampling methods applied during the comprehensive air toxics assessment at LGTI. In some cases, syngas methods were modified to account for the nature and reactivity of the syngas components. Sampling for hydrogen cyanide is one example of a method modification. A buffered lead acetate solution (pH 4) was used prior to the HCN collecting impingers to remove H_2S by precipitation. Zinc acetate was used as the collecting agent instead of NaOH to eliminate the problems associated with the absorption of

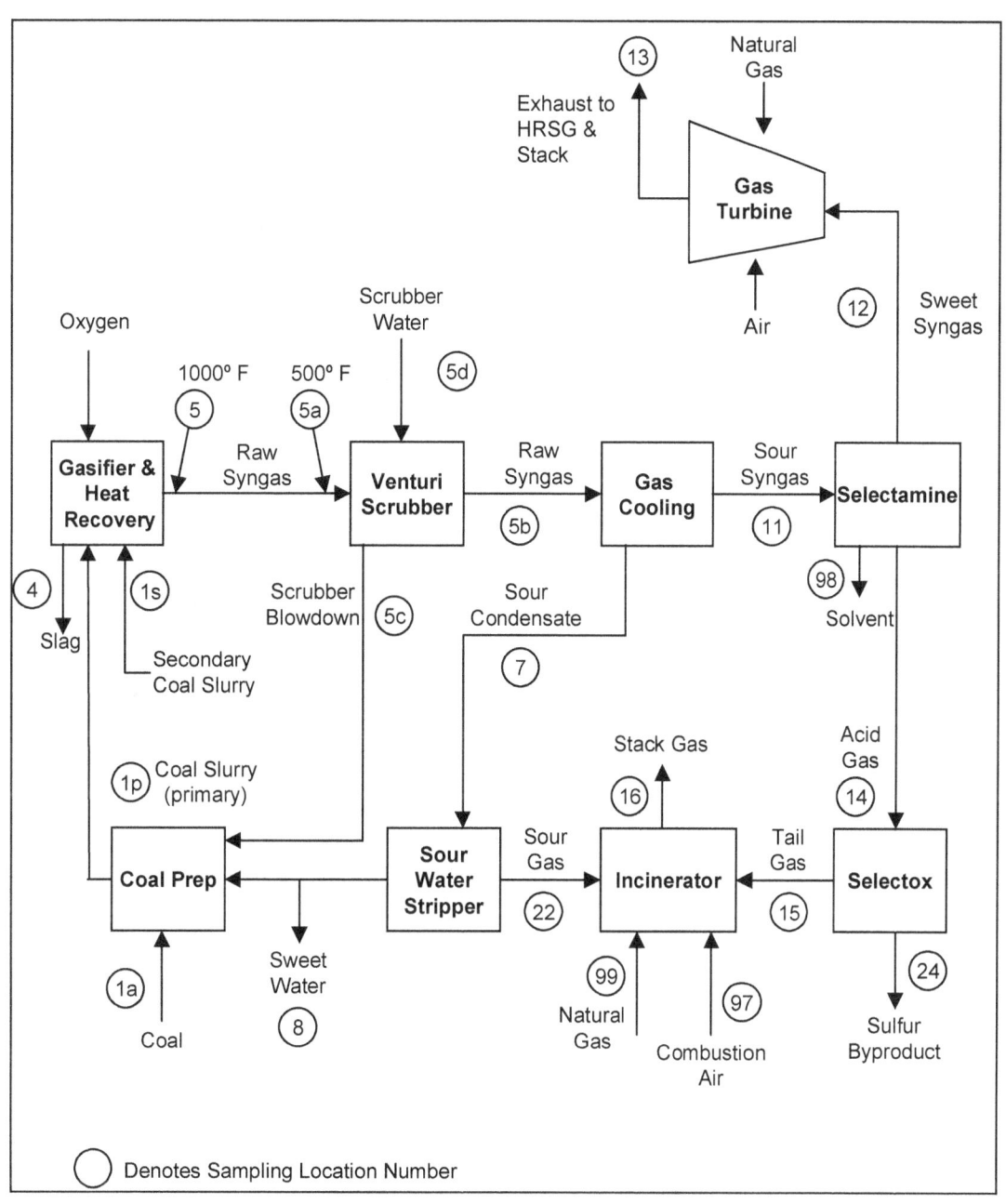

Figure 5-3. Sampling Locations for Comprehensive Testing at LGTI

Table 5-9. Sampling Locations and Analytes

Location	Stream	Test Period	Analytes
1a	Coal pile	1, 2, 3	Metals, ultimate, proximate, anions
		3	Radionuclides
1p, 1s	Coal slurry	1, 2, 3	Metals, ultimate, proximate, anions
4	Slag	1, 2, 3	Metals, ultimate, proximate, anions
		3	Radionuclides
5	Raw gas, 1,000• F	4	Vapor: metals, Cl, F, NH_3, HCN Particulate: metals
5a	Raw gas, 500• F	3	Metals, C_1-C_{10}, Cl, F, NH_3, HCN
5a	Raw gas, 500• F	probe shakedown test	Particulate: metals
5b	Raw gas, scrubbed	3	Metals, C_1-C_{10}, Cl, F, NH_3, HCN
5c	Scrubber blowdown (char)	3	Metals, ultimate, proximate, anions
	(filtrate)	3	Metals, ultimate, proximate, anions, ammonia, cyanide, suspended solids
5d	Scrubber water	3	Metals, ultimate, proximate, anions, ammonia, cyanide
7	Sour condensate	2	Metals, cyanide, volatile/semi-volatile organics, aldehydes, anions, ammonia, phenol, sulfide, water quality
8	Sweet water	2	Metals, cyanide, volatile/semi-volatile organics, aldehydes, anions, ammonia, phenol, sulfide, water quality
11	Sour syngas	1	Particulates, metals, C_1-C_{10}, volatile organics, major gases, sulfur species, semi-volatile organics, aldehydes, Cl, F, NH_3, HCN
12	Sweet syngas	1	Particulates, metals, C_1-C_{10}, volatile organics, major gases, sulfur species, semi-volatile organics, aldehydes, Cl, F, NH_3, HCN
13	Turbine Exhaust	1	Particulates, PM-10, metals, VOST, semi-volatile organics, aldehydes, Cl, F, NH_3, HCN, H_2SO_4, CEM gases
14	Acid gas	1	Metals, C_1-C_{10}, major gases, sulfur species, semi-volatile organics, Cl, F, NH_3, HCN
15	Tail gas	1	Metals, C_1-C_{10}, major gases, sulfur species, semi-volatile organics, NH_3, HCN
		2	C_1-C_{10}, sulfur species, semi-volatile organics, NH_3, HCN CEM gases
16	Incinerator stack	2	Particulates, PM-10, metals, VOST, sulfur species, semi-volatile organics, aldehydes, Cl, F, NH_3, HCN, H_2SO_4, CEM gases
22	Sour gas	2	C_1-C_{10}, major gases, NH_3, HCN
24	Sulfur	1	Metals, ultimate, proximate
97	Combustion air	2	C_1-C_{10}, major gases, sulfur species, NH_3, HCN
98	Selectamine ™ solvent	1	Ash, volatile organics, heat stable salts
		3	Ash, heat stable salts
99	Natural gas	2	Metals, C_1-C_{10}, sulfur species

Table 5-10. Summary of Sampling Methods for Syngas

Analyte	Sampling Method	Preparation and Analytical Methods	Comments on Applicability
Particulate Loading	EPA Method 5 (adapted)	Method 5 - Gravimetric	A specialized sampling system must be designed to extract an isokinetic sample from a pressurized, and flammable gas stream.
Metals (Particulate phase)	EPA Method 29 (SW0060)	EPA Method 29, SW 0060, SW 6010, SW 7470	
Metals (Vapor Phase)	EPA Method 29 (SW 0060) - modified	EPA Method 29, SW 0060, SW 6010, SW 7470	Ineffective for most metals in syngas matrix (i.e., poor collection efficiency). HNO₃/H₂O₂ impinger solution concentrations were increased to improve oxidation potential - this also proved ineffective at enhancing collection efficiency. The use of KMnO₄/H₂SO₄ solutions for Hg collection were omitted - incompatible with reduced gases (H₂S).
	Charcoal Adsorption (Radian method)	HNO₃ acid digestion; SW 0060, SW 6010, SW 7470	Appears to be applicable for mercury and possibly other vapor phase trace elements, however this application has not been validated. NOTE – Based on further developments, sample dissolution by combustion of charcoal adsorbent in a closed oxygen combustion bomb (ASTM D3684) is now recommended.
	Continuous on-line AAS analyzer (Radian method)	Direct analysis by atomic absorption spectrometry (AAS)	Limited to single element analysis (1 element per hour), but theoretically provides the most definitive results.
Mercury (Vapor phase)	EPA Method 101 (adapted)	SW 7470	NaOH impingers added to sampling train prior to KMnO₄/H₂SO₄ impingers for removal of H₂S. All impinger solutions analyzed for Hg.
Ammonia	EPA Method 26 (SW 0050) - modified	EPA Method 350.2/350.1	Increasing the sulfuric acid impinger solution concentration is necessary for high-ammonia streams.
Halides/Acid gases		SW 9057 (chloride)	This approach was performed in lieu of alkaline carbonate impinger solutions to reduce the number of sampling trains required during the tests. Detection limits for chloride are higher due to the high concentration of sulfuric acid and its effect on the ion chromatography method. The use of a dilute bicarbonate/carbonate solution for halides and acid gases is recommended for increased analytical sensitivity.
		EPA 340.2 (fluoride)	
Hydrogen Cyanide	Texas Air Control Board Method - modified by Radian	SW 9012	Buffered lead acetate solution used prior to collecting impingers to remove H₂S. Zinc acetate solution used to collect cyanide. Due to the level of CO₂ in syngas, the use of NaOH solutions creates a negative bias through several mechanisms affecting both collection and analysis.

Table 5-10 (continued)

Analyte	Sampling Method	Preparation and Analytical Methods	Comments on Applicability
Semi-volatile Organic Compounds	SW 0010	SW 8270	This method was not performed with any pre-sampling surrogate spikes on the XAD resin to assess retention of SVOCs during sampling. The effect of syngas and other internal process stream components has not been determined. Post sampling surrogate spikes were recovered within method specified objectives indicating acceptable analytical performance.
Dioxins and Furans	EPA Method 23	EPA Method 23 SW 8290	Acceptable pre-sampling surrogate spike recoveries have been reported.
Aldehydes	SW 0011	SW 0011	This method was not performed with any pre-sampling spikes to assess retention of aldehydes during sampling. The effect of syngas and other internal process stream components has not been determined. Post sampling surrogate spikes were recovered within method specified objectives indicating acceptable analytical performance.
Volatile Organics	Summa Canister	TO 14 (GC/MS)	VOST (SW0030) not applicable due to absorption of H_2S on charcoal adsorbent tubes.
Major Gas Composition	Pressurized gas sample cylinder or Tedlar bag	GC/TCD	
Reduced sulfur compounds		GC/FPD	On-site analysis required unless samples are collected in Teflon-lined, or aluminum cylinders to prevent loss of reduced sulfur compounds by passivation of steel bomb components.

CO_2. When CO_2 is absorbed in NaOH, the pH of the solution is reduced and this results in poor collection efficiency as well as a low analytical bias from the distillation of CO_2 from the sample.

Characterization of vapor-phase metals is the least developed of all methods for syngas. EPA Method 29 (SW-846 Method 0060) has been shown ineffective for the collection of vapor phase metals in syngas. Charcoal adsorbents have been tested and have demonstrated some success, but selectivity in collection may prohibit their use for characterizing a full suite of elements. Other direct measurement techniques have also been applied such as atomic absorption spectrometry, plasma emission spectrometry, and flame emission spectrometry. These methods replace or supplement the fuel gas (acetylene, hydrogen) or plasma gas to the instrument with the syngas sample. Any metals present in the gas will be atomized in the flame or plasma for measurement by atomic absorption or emission. In theory, this provides a total measurement of the elements present, regardless of their form. Unfortunately, they are expensive and time consuming to operate and detection limits are seldom low enough for a complete characterization of the syngas.

Charcoal adsorbents offer the best approach available at this time, however further development is needed. The US Department of Energy has sponsored a sampling and analytical method development project to address this issue, specifically for the analysis of mercury in syngas (EG&G subcontract #721041, release #825551). Radian International was contracted to perform this method development based on the method performance data gathered during the LGTI gasifier tests. At this time, two viable methods for mercury have been successfully tested in a bench-scale laboratory study; an impinger method (Hg only) and a charcoal adsorbent method (possible extension to other metals). Field testing of the methods for validation has not been performed pending the selection of a suitable test site.

5.4   Conclusions

Both gasification and incineration are capable of converting hydrocarbon-based hazardous materials to simple, nonhazardous byproducts. However, the conversion mechanisms and the nature of the byproducts differ considerably and these factors should justify the separate treatment of these two technologies in the context of environmental protection and economics.

Gasification technologies meeting the definition proposed by the GTC offer an alternative process for the recovery and recycling of low-value materials by producing a more valuable commodity - syngas. The multiple uses of syngas (power production, chemicals, methanol, etc.) and the availability of gas cleanup technologies common to the petroleum refining industry make gasification of secondary oil-bearing materials a valuable process in the extraction of products from petroleum. By producing syngas, sulfur, and metal-bearing slag suitable for reclamation, wastes are minimized and the emissions associated with their destruction by incineration are reduced.

Data on syngas composition from the gasification of a wide variety of feedstocks (oil, petroleum coke, coal, and various hazardous waste blends) indicates the major components of syngas to consistently be CO, H_2, and CO_2 with low levels of N_2 and CH_4 also present. Hydrogen sulfide levels in the raw syngas are related to the sulfur content of the feedstock. Similarly, NH_3 and HCN concentrations are related to the fuel's nitrogen content, and HCl levels are affected by the fuel's chlorine content.

Organic compounds such as benzene, toluene, naphthalene, and acenaphthalene have been detected at very low levels in the syngas from some gasification systems. However, when used as a fuel and combusted in a gas turbine, the emissions of these compounds or other organic HAPs are either not detected or present at sub-part-per-billion concentrations in the emitted stack gas. In addition, emissions of particulate matter are found to be one to two orders of magnitude below the current RCRA emissions standards and the recently proposed MACT standard for hazardous waste incinerators.

Although comprehensive test data from the gasification of coal and other fossil fuels are available to assess the fate of many hazardous constituents, the same type and volume of data for the gasification of hazardous wastes are not readily available. To fully assess the performance of gasification on a broader spectrum of hazardous wastes, additional testing may be required to fill data gaps and provide validation of test methods.

All things considered, the ability of gasification technologies to extract useful products from secondary oil-bearing materials and listed refinery wastes is analogous to petroleum coking operations and unlike hazardous waste incineration. Like petroleum coking, gasification can be viewed as an integral part of the refining process where secondary oil-bearing materials can be converted to a fuel (syngas) that is of comparable quality to the syngas produced from the gasification of fossil fuels.

5.5   References

1. U.S. EPA Final Rule 63 FR 42110, August 6, 1998.

2. U.S. EPA Proposed MACT Rule for Hazardous Waste Combustors, 61 FR 17357, April 19,1996.

3. Oppelt, T.E. "Incineration of Hazardous Waste: A Critical Review," *JAPCA*, Volume 37, No. 5, May 1987.

4. Dempsey, C., and Oppelt, T.E. "Incineration of Hazardous Waste: A Critical Review Update," *Air and Waste*, Vol. 43, January 1993.

5. Heaven, D.L, "Gasification Converts a Variety of Problem Feedstocks and Wastes," *Oil & Gas Journal*, May 1996.

6. Electric Power Research Institute, "Cool Water Coal Gasification Program: Final Report," prepared by Radian Corporation and Cool Water Coal Gasification Program. EPRI Final Report GS-6806, December 1990.

7. U.S. EPA, "Texaco Gasification Process Innovative Technology Evaluation Report," Office of Research and Development Superfund Innovative Technology Evaluation Program, EPA/540/R-94/514, July 1995.

8. U.S. EPA, "Draft Technical Support Document for HWC MACT Standards Volume II: HWC Emissions Database," Office of Solids Waste and Emergency Response, February 1996.

9. Tuppurainen, K., I. Halonen, P. Ruokojarvi, J. Tarhanen, and J. Ruuskanen. "Formation of PCDDs and PCDFs in Municipal Waste Incineration and Its Inhibition Mechanisms: A Review," University of Kuopio, Kuopio, Finland, *Chemosphere* (in press).

10. Halonen, I. "Formation and Prevention of Polychlorinated Dibenzo-p-dioxins and Dibenzofurans in Incineration Processes," University of Kuopio, Kuopio, Finland. Doctoral Dissertation, November 1997.

11. Huang, H. and A. Buekens. "On the Mechanisms of Dioxin Formation in Combustion Processes," *Chemosphere*, Vol. 32, No. 9, pp. 4099-4117, November 1995.

12. Gullett, B.K. and K. Raghunathan. "Observations on the Effect of Process Parameters on Dioxin/Furan Yield in Municipal Waste and Coal Systems," *Chemosphere*, Vol. 34, No. 5-7, pp. 1027-1032, March - April 1997.

13. Raghunathan, K. and B.K. Gullett. "Role of Sulfur in Reducing PCDD and PCDF Formation," *Environmental Science and Technology*, Vol. 30:6, pp. 1827-1834, June 1996.

14. Halonen, I., J. Tarhanen, P. Ruokojarvi, K. Tuppurainen, and J. Ruuskanen, "Effects of Catalysys and Chlorine Source on the Formation of Organic Chlorinated Compounds, *Chemosphere*, Vol. 30, No. 7, pp. 1261-1273, 1995.

15. Addink, R., A.J. Harrie, Grovers, and Kees Olie, "Kinetics of Formation of Polychlorinated Dibenzo-p-dioxins/Dibenzofurans from Carbon on Fly Ash," *Chemosphere*, Vol. 31, No. 6, pp. 3549-3552, September 1995

16. Chagger, H.K., J.M. Jones, M. Pourkashanian, and A. Williams, "The Nature of Hydrocarbon Emissions Formed during the Cooling of Combustion Products," *Fuel*, Vol 76, No. 9, pp. 861-864, July 1997.

17. Ghorishi, S.B. and E.R. Altwicker, "Rapid High Temperature Formation of Polychlorinated Dioxins and Furans in the Bed Region of a Heterogeneous Spouted Bed Combustor: Development of a Surface Mediated Model for the Formation of Dioxins," *Hazardous Waste and Hazardous Materials*, Vol. 1, No. 1, pp. 11-12, Spring 1996.

18. Cains, P.W., L.J. McCausland, A.R. Fernandes, and P. Duke, "Polychlorinated Dibenzo-p-dioxins and Dibenzofurans Formation in Incineration: Effects of Fly Ash and Carbon Source," *Environmental Science and Technology*, Vol. 31, No. 3, pp. 776-785, 1997.

19. Electric Power Research Institute, *Summary Report: Trace Substance Emissions from a Coal-Fired Gasification Plant*, prepared for EPRI and the U.S. Department of Energy, June 29, 1998.

20. The Hazardous Waste Consultant. Dioxins—Their Measurement and Control, September/October 1997.

21. Gasification Technology Counsel. Response to Comments in ETC letter of October 13, 1998 and EDF letter of October 13, 1998. Letter for RCRA Docket Number F-98-PR2A-FFFFF, May 13, 1998.

22. U.S. EPA, "Performance Evaluation of Full-Scale Hazardous Waste Incineration," five volumes, NTIS, PB- 85-129500, November 1994.

23. D. Van Buren, G. Pie, C. Castaldini, "Characterization of Hazardous Waste Incineration Residuals," U.S. EPA, January 1987.

24. U.S. EPA, Notice of Data Availability (NODA) Response to Comment Document: Part II, Office of Solid Waste, June 1998.

25. Electric Power Research Institute. "Coal Gasification Guidebook: Status, Applications and Technologies". EPRI TR-102034, p. 5-48, 5-58. 1993.

26. Rath, "Status of Gasification Demonstration Plants," Proc. 2nd Annual Fuel Cells Contract Review Meeting. DOE/METC-9090/6112, p. 91.

27. U.S. EPA. Final MACT Rule for Hazardous Waste Combustors, 64 FR 52828, September 30, 1999.

28. Baker, D.C., "Projected Emissions of Hazardous Air Pollutants from a Shell Coal Gasification Process-Combined-Cycle Power Plant," *Fuel*, Vol. 73, No. 7, 1994.

29. DelGrego, G., "Experience with Low Value Feed Gasification at the El Dorado, Kansas Refinery," Presented at the 1999 Gasification Technologies Conference, San Francisco, CA, October 17-20, 1999.

30. Liebner, W., "MGP-Lurgi/SVZ Mulit Purpose Gasification, Another Commercially Proven Gasification Technology," Presented at the 1999 Gasification Technologies Conference, San Francisco, CA, October 17-20, 1999.

31. De Graaf, J.D., E.W. Koopmann, and P.L. Zuideveld, "Shell Pernis Netherlands Refinery Residue Gasification Project," Presented at the 1999 Gasification Technologies Conference, San Francisco, CA, October 17-20, 1999.

32. Collodi, G. and R.M. Jones, "The Sarlux IGCC Project and Outline of the Construction and Commissioning Activities," Presented at the 1999 Gasification Technologies Conference, San Francisco, CA, October 17-20, 1999.

33. Skinner, F.D., *Comparison of Global Energy Slagging Gasification Process for Waste Utilization with Conventional Incineration Technologies*. Final Report, Radian Corporation, January 1990.

34. Vick, S.C., "Slagging Gasification Injection Technology for Industrial Waste Elimination," Presented at the 1996 Gasification Technologies Conference, San Francisco, CA, October 1996.

35. Seifert, W., "Utilization of Wastes – Raw Materials for Chemistry and Energy. A Short Description of the SVZ-Technology," Prepared for the technical conference: "Gasification the Gateway to a Cleaner Future" Dresden, Germany, September 23-24, 1998.

36. Salinas, L., P. Bork and E. Timm, "Gasification of Chlorinated Feeds," Presented at the 1999 Gasification Technologies Conference, San Francisco, CA, October 17-20, 1999.

37. The Thermoselect Solid Waste Treatment Process. Vendor literature supplied by Thermoselect Incorporated. Troy, Michigan, 1999.

38. Keeler, C.G., "Wabash River in Its Fourth Year of Commercial Operation," Presented at the 1999 Gasification Technologies Conference, San Francisco, CA, October 17-20, 1999.